International Federation of Automatic Control

DISTRIBUTED COMPUTER CONTROL SYSTEMS

Other Titles in the IFAC Proceedings Series

ATHERTON: Multivariable Technological Systems
BANKS & PRITCHARD: Control of Distributed Parameter Systems
CICHOCKI & STRASZAK: Systems Analysis Applications to Complex Programs
CRONHJORT: Real Time Programming 1978
CUENOD: Computer Aided Design of Control Systems
De GIORGO & ROVEDA: Criteria for Selecting Appropriate Technologies under Different Cultural, Technical and Social Conditions
DUBUISSON: Information and Systems
GHONAIMY: Systems Approach for Development
HASEGAWA & INOUE: Urban, Regional and National Planning - Environmental Aspects
ISERMANN: Identification and System Parameter Estimation
LAUBER: Safety of Computer Control Systems
LEONHARD: Control in Power Electronics and Electrical Drives
MUNDAY: Automatic Control in Space
NIEMI: A Link Between Science and Applications of Automatic Control
NOVAK: Software for Computer Control
OSHIMA: Information Control Problems in Manufacturing Technology (1977)
REMBOLD: Information Control Problems in Manufacturing Technology (1979)
RIJNSDORP: Case Studies in Automation related to Humanization of Work
SAWARAGI & AKASHI: Environmental Systems Planning, Design and Control
SINGH & TITLI: Control and Management of Integrated Industrial Complexes
SMEDEMA: Real Time Programming 1977
TOMOV: Optimization Methods - Applied Aspects

NOTICE TO READERS

Dear Reader

If your library is not already a standing order customer or subscriber to this series, may we recommend that you place a standing or subscription order to receive immediately upon publication all new volumes published in this valuable series. Should you find that these volumes no longer serve your needs your order can be cancelled at any time without notice.

ROBERT MAXWELL
Publisher at Pergamon Press

DISTRIBUTED COMPUTER CONTROL SYSTEMS

*Proceedings of the IFAC Workshop,
Tampa, Florida, U.S.A., 2-4 October 1979*

Edited by

T. J. HARRISON

*International Business Machines Corporation,
Armonk, New York, U.S.A.*

Published for the

INTERNATIONAL FEDERATION OF AUTOMATIC CONTROL

by

PERGAMON PRESS

OXFORD · NEW YORK · TORONTO · SYDNEY · PARIS · FRANKFURT

U.K.	Pergamon Press Ltd., Headington Hill Hall, Oxford OX3 0BW, England
U.S.A.	Pergamon Press Inc., Maxwell House, Fairview Park, Elmsford, New York 10523, U.S.A.
CANADA	Pergamon of Canada, Suite 104, 150 Consumers Road, Willowdale, Ontario M2J 1P9, Canada
AUSTRALIA	Pergamon Press (Aust.) Pty. Ltd., P.O. Box 544, Potts Point, N.S.W. 2011, Australia
FRANCE	Pergamon Press SARL, 24 rue des Ecoles, 75240 Paris, Cedex 05, France
FEDERAL REPUBLIC OF GERMANY	Pergamon Press GmbH, 6242 Kronberg-Taunus, Hammerweg 6, Federal Republic of Germany

Copyright © IFAC 1980

All Rights Reserved. No part of this publication may be reproduced, stored in a retrieval system or transmitted in any form or by any means: electronic, electrostatic, magnetic tape, mechanical, photocopying, recording or otherwise, without permission in writing from the copyright holders.

First edition 1980

British Library Cataloguing in Publication Data

IFAC Workshop on Distributed Computer Control
Systems, *Florida, 1979*
Distributed computer control systems.
1. Automatic control - Data processing - Congresses
2. Electronic data processing - Distributed
processing - Congresses
I. Title   II. Harrison, Thomas J
III. Instrument Society of America   IV. American
Automatic Control Council   V. International
Federation of Automation Control
629.8'95   TJ212   80-40561

ISBN 0 08 024490 4

These proceedings were reproduced by means of the photo-offset process using the manuscripts supplied by the authors of the different papers. The manuscripts have been typed using different typewriters and typefaces. The lay-out, figures and tables of some papers did not agree completely with the standard requirements; consequently the reproduction does not display complete uniformity. To ensure rapid publication this discrepancy could not be changed; nor could the English be checked completely. Therefore, the readers are asked to excuse any deficiencies of this publication which may be due to the above mentioned reasons.

The Editor

Printed in Great Britain by A. Wheaton & Co. Ltd., Exeter

IFAC WORKSHOP ON DISTRIBUTED COMPUTER CONTROL SYSTEMS

Organized by
The Instrument Society of America
for
The American Automatic Control Council

Sponsored by
The International Federation of Automatic Control Computer Committee (COMPCON)

International Program Committee
T. J. Harrison, U.S.A. (Chairman)
P. F. Elzer, F.R.G.
R. W. Gellie, Canada
R. C. Jaeger, U.S.A.
K. D. Müller, F.R.G.
T. Tohyama, Japan

CONTENTS

List of Participants — ix

Preface — xi

Welcoming Statement
C. M. Doolittle — xiii

Session 1.   R. W. Gellie, Chairman

The ISABELLE Control System – Design Concepts
J. W. Humphrey — 1

Discussion — 17

A Distributed Control System for Facility Energy Management
C. A. Garcia and V. A. Kaiser — 21

Discussion — 35

Session 2.   R. C. Jaeger, Chairman

Programming Distributed Computer Systems with Higher Level Languages
H. U. Steusloff — 39

Discussion — 51

A Distributed Control System – Concept and Architecture
S. Yoshii and S. Kuroda — 53

Discussion — 61

Structure of an Ideal Distributed Computer Control System
Th. Lalive d'Epinay — 65

Discussion — 71

Session 3.   T. J. Harrison, Chairman

Panel Discussion

Distributed Data Processing Systems—Distributed Computer Control Systems: Similarities and Differences
T. J. Harrison — 75

Introductory Statements:
T. L. Johnson — 81
W. L. Miller — 83
R. P. Singer — 87
H. U. Steusloff — 91

Discussion — 95

Session 4. W. L. Miller, Chairman

A Multi-processor Approach for the Automation of Quality Control in an Overall Production Control System
K. Zwoll, K. D. Müller and J. Schmidt — 105

Concept and Examples of a Distributed Computer System with a High-Speed Ring Data Bus
T. Kawabata and H. Tanaka — 119

Discussion — 131

Session 5. F. G. Sheane, Chairman

Transaction Processing in Distributed Control Systems
R. G. Wilhelm, Jr. — 133

Discussion — 141

Parallel Processing Concepts for Distributed Computer Control Systems
D. AlDabass — 143

Discussion — 155

The Design of Distributed Computer Control Systems
A. Nader — 157

Discussion — 181

Session 6. K. D. Müller, Chairman

Development of Distributed Control Systems
F. J. Romeu — 183

Discussion — 193

Additional Papers

Petri Nets for Real Time Control Algorithms Decomposition
A. Nader — 197

Economic Considerations for Real-Time Aircraft/Avionic Distributed Computer Control Systems
B. A. Zempolich — 211

Author Index — 229

LIST OF PARTICIPANTS

IFAC WORKSHOP ON DISTRIBUTED COMPUTER CONTROL SYSTEMS

Host International Hotel, Tampa, Florida, U.S.A.

2-4 October 1979

*Dr. David AlDabass
University of Manchester
Institute of Science and Technology
P. O. Box 88
Manchester M60 1QD
ENGLAND

Mr. Jude Anders
Johnson Controls, Inc.
P. O. Box 423
Milwaukee, WI 53201, USA

Dr. Richard C. Born
Cutler-Hammer
4201 North 27 Street
Milwaukee, WI 53216, USA

Dr. Marcus R. Buchner
Assistant Professor
Case Western Reserve University
Systems & Computer Engineering
Crawford Hall, Room 608
10900 Euclid Avenue
Cleveland, OH 44106, USA

Mr. Richard H. Caro
MODCOMP
P. O. Box 6099
Fort Lauderdale, FL 33310, USA

Mr. Thomas Casey
Modular Computer Systems, Inc.
1650 West McNab Road
Fort Lauderdale, FL 33310, USA

Mr. Renzo Dallimonti
Honeywell
1100 Virginia Drive
Fort Washington, PA 19034, USA

*Dr. T. Lalive D'Epinay
Head of Automatic Control
Brown Boveri Research Centre
CH-5405 Baden
SWITZERLAND

Mr. Charles M. Doolittle
IBM Corporation
1133 Westchester Avenue
White Plains, NY 10604, USA

* = Author
** = Program/Organizing Committee

Ms. Evelyn Dow
Gould, Inc.
Modicon Division
P. O. Box 83, S.V.S.
Andover, MA 01810, USA

*Mr. Carlos Garcia
IBM Corporation
1000 Westchester Avenue
White Plains, NY 10604, USA

**Dr. R. W. Gellie
National Research Council
  of Canada
Building M3
Ottawa, K1A0R6
CANADA

**Dr. Thomas J. Harrison
IBM Corporation
P. O. Box 1328
Boca Raton, FL 33432, USA

*Dr. J. W. Humphrey
Assistant Accelerator Division Head
Isabelle Project
Brookhaven National Laboratory
Upton, NY 11973, USA

Mr. Ben Hyde
Bolt-Beranek and Newman, Inc.
50 Maulton Street
Cambridge, MA 02138, USA

**Dr. R. C. Jaeger
Associate Professor
Dept. of Electrical Engineering
Dunstan Hall
Auburn University
Auburn, AL 36830, USA

*Dr. Timothy L. Johnson
Associate Professor EE & CS
Laboratory for Information and
  Decision Systems
RM 35-210
Massachusetts Institute of
  Technology
Cambridge, MA 02139, USA

*Mr. Vaughn A. Kaiser
Profimatics, Inc.
6355 Topanga Canyon Boulevard
Woodland Hills, CA 91367, USA

List of Participants

Mr. Simon Korowitz
Leeds and Northrup
Sumneytown Pike
North Wales, PA 19454, USA

*Mr. Tetsuo Kawabata
Toshiba Corporation
Toshiba Fuchu Works
1, Toshiba-Cho, Fuchu City
Tokyo, 183
JAPAN

Mr. W. E. Klein
Sperry Univac
P. O. Box 500
Blue Bell, PA 19424, USA

*Mr. Seinosuke Kuroda
Hokushin Electric Works, Ltd.
Shuwa Onarimon Building No. 1
6-Chome, Shimbashi
Minato-ku, Tokyo 105
JAPAN

**Mr. Donald L. Lee
IBM Corporation
P. O. Box 1328
Boca Raton, FL 33432, USA

Mr. Leo J. Lambert
General Electric SPCC
Bldg. 31 EE 1285
Boston Avenue
Bridgeport, CT 06602, USA

Mr. David L. Legrow
Scott Paper Company
1 Scott Plaza
Philadelphia, PA 19113, USA

Dr. Paul Merluzzi
Stauffer Chemical Co.
Dobbs Ferry, NY 10522, USA

*Dr. William L. Miller
Fischer and Porter
Department 439
East Country Line Road
Westminster, PA 18974, USA

Mr. Jerry A. Moon
Fisher Controls
Marshalltown, IA 50158, USA

*Dr. Klaus D. Müller
**Scientist
Kernforschungsanlage (KFA) Julich GmbH
P. O. Box 1913
5170 Julich
WEST GERMANY

Professor G. Musstopf
Fachbereich Informatik
Hamburg University
Schlueterstr. 70
D-2000 Hamburg 13
WEST GERMANY

Mr. D. P. Reed
Supervisor, Process Computers
Kaiser Steel Corporation, A-414
P. O. Box 217
Fontana, CA 92335, USA

*Mr. Fred G. Sheane
Control Technology Group
Imperial Oil Ltd.
P. O. Box 3004
Sarnia, Ontario N7T 7M5
CANADA

Mr. Rubin P. Singer
IBM Corporation, 1-A
1000 Westchester Avenue
White Plains, NY 10604, USA

Mr. Richard W. Signor
General Electric Co.
Bldg. 31EE
1285 Boston Avenue
Bridgeport, CT 06602, USA

*Dr. Hartwig U. Steusloff
Fraunhofer Gesellschaft
Fraunhofer Institute for Information
   and Data Processing
Sebastian-Kneipp Strasse 12-14
D-7500 Karlsruhe 1
WEST GERMANY

*Mr. Robert G. Wilhelm, Jr.
Industrial Nucleonics Corporation
650 Ackerman Road
Columbus, OH 43202, USA

Mr. Robert Willard
Digital Equipment Corporation
146 Main Street
Maynard, MA 01754, USA

*Dr. Seiji Yoshii
Hokushin Electric Works, Ltd.
3-30-1, Shimomaruko Ohta-ku
Tokyo 146
JAPAN

Mr. Roy W. Yunker
Director, Process Control
P. P. G. Industries
One Gateway Center
Pittsburgh, PA 15222, USA

*Mr. Frank J. Romeu
Process Management Systems Division
Honeywell Inc.
1100 Virginia Drive
Fort Washington, PA 19034, USA

Mr. T. B. Rooney
The Foxboro Company
Neponset Avenue
Foxboro, MA 02035, USA

Dr. H. Sandmayr
BBC Brown, Boveri & Company, Limited
CH-5405 Baden-Dättwil
SWITZERLAND

*Dr. H. K. Zwoll
Scientist
Kerforchungsanlage (KFA) Julich GmbH
P. O. Box 1913
5170 Julich
WEST GERMANY

PREFACE

The first IFAC Workshop on Distributed Computer Control Systems was held in Tampa, Florida, USA, on October 2-4, 1979. This Workshop, and the plan to make them an annual event, was the idea of C. M. Doolittle, Chairman of COMPCOM. Charlie provides some insight into his reasoning in the introduction which he provided for this volume.

In agreeing to organize the first of these workshops, I also had several ideas of what should be accomplished. Distributed computing is a timely topic and I felt it important that the papers presented represent new ideas, not previously published. As a result, and in accordance with IFAC guidelines, papers were selected on the basis of short abstracts and the final papers were not available until the time of the Workshop. This eliminated the delay which often comes between the writing and the publishing of a paper.

It also seemed important to me that the participants at the Workshop have sufficient time to thoroughly discuss each of the papers. For this reason, at least a full hour was devoted to the paper presentation and the discussion. The discussions often continued into the lunch hour and coffee breaks as small groups of participants engaged in active interchange of ideas and opinions. This interchange was, I believe, facilitated by two factors: the group was reasonably small (43 participants from six countries), and some participants knew each other from previous IFAC and ISA meetings. These factors did much to minimize the usual reluctance to become involved in the discussion. As a result, it was a week full of active and animated discussion by all of the participants.

My third thought in organizing the Workshop stemmed from my observation that there seems to be little interaction between computer users and vendors involved in "classic data processing" and those involved with the use of computers in industrial processes.

Therefore, a specific effort was made to attract people from both the data processing and industrial control subspecialties of the computer field. In reviewing the papers, the backgrounds of the authors, and the affiliations of the other participants, you will find that this goal was realized. The papers span a spectrum from theoretical control theory to the practical aspects of managing the process by which control computers are designed and built.

The Workshop was divided into six sessions, each led by a Session Chairman. The papers selected for each session were chosen to foster discussion, although session themes were not specifically identified. In these proceedings, the papers are published in the order in which they were presented. It will be noted that this temporal order makes some difference in understanding the discussion transcripts.

The discussions following each paper and the panel discussion were recorded and transcibed in their entirety. Some editing of the transcript has been done to clarify portions of the discussion and to eliminate obvious redundancy. The editing was not extreme, however, and some redundancy remains. This is intentional in that I feel that the reasonable complete transcript provides a flavor of the Workshop which would have been lost through an abnormally heavy editor's hand.

The discussions were recorded using conventional microcassette dictation equipment and, despite our training in process control, we failed to have a complete feedback loop which tested to make sure that what was being heard in our monitor headsets was, in fact, being recorded on the tape. We later found, to our dismay, that in several cases we had electronic failures and the recorded material was obliterated by uncontrolled oscillations. In one case, we were able to reconstruct the discussion from notes. In the case of Dr. Müller's paper and for a part of the discussion of Mr. Romeu's paper, we were unable to meaningfully reconstruct the discussion. We apologize for this poor engineering and will pass a suggested design on to the next Workshop organizer to, hopefully, avoid similar problems at the next Workshop.

Several other notes on the published papers are in order. Dr. Nader was not able to attend the Workshop. He had contributed two papers: One was ably presented by Dr. David AlDabass. The other, concerning the use of Petri Nets, was not presented but is provided as part of these proceedings. It was also not possible for Mr. Zempolich to attend the Workshop. His paper is included here, although it was not presented at the Workshop.

The organization of this Workshop, like all others, involved the time, effort, and interest of many people. I would particularly like to thank Mr. Don Lee, who was the Arrangements Chairman for the Workshop, for his many hours of effort which contributed to a smooth running meeting. The International Program Committee, listed in the front pages of this volume, provided advice, counsel, and assistance in finding and selecting the papers which were presented. A number of organizations contributed through providing us skilled personnel and/or equipment: The Tampa Chamber of Commerce assisted in the registration process and provided local information; the staff of the Instrument Society of America (ISA) prepared and mailed the call for papers, and furnished artwork, and other material; the University of South Florida provided student volunteers during several of the sessions; and Hobart Business Machines of Boca Raton, Florida, the IBM Corporation, and Xerox Corporation provided office equipment and other supplies for our use. Without the efforts of these people and organizations, the Workshop would not have been nearly as successful.

Thomas J. Harrison
Workshop Chairman

December 31, 1979
Armonk, N.Y., U.S.A.

WELCOME TO THE FIRST IFAC WORKSHOP ON DISTRIBUTED COMPUTER CONTROL SYSTEMS

C. M. Doolittle

IBM Data Processing Division, 1133 Westchester Avenue, White Plains, NY 10604, U.S.A.

The IFAC Technical Committee on Computers extends a warm welcome to those attending the first IFAC Workshop on Distributed Computer Control Systems.

We expect to make this workshop an annual event because of the significance of the subject and the current level of interest. We count on our experience in other similar endeavors to guide us. For example, the ninth annual IFAC Workshop on Real Time Programming was sponsored this year by the Computers Committee. And, the second triennial Symposium on Software for Computer Control was also held this year in Prague, sponsored by our committee.

The purpose of IFAC is to further the science, technology and application of automatic control in the broadest sense in all systems. It has been doing that now for 22 years through its congresses, symposia, workshops and publications.

The real machinery of IFAC is the voluntary efforts of people interested in control, in learning more, in sharing what they have learned and in using the processes offered by an international professional society federation.

Organizers of these events do a lot of extra work and take some risks to develop these meetings and make them successful.

IFAC is especially grateful for the efforts of Tom Harrison and his program committee. They have given us a way to share and expand our knowledge in an important area and the Computers Committee plans to continue in the same

THE ISABELLE CONTROL SYSTEM—
DESIGN CONCEPTS*

J. W. Humphrey

Brookhaven National Laboratory, Upton, NY 11973, U.S.A.

Abstract. ISABELLE is a Department of Energy funded proton accelerator/storage ring being built at Brookhaven National Laboratory (Upton, Long Island, New York). It is large (3.8 km circumference) and complicated (~ 30 000 monitor and control variables). It is based on superconducting technology. Following the example of previous accelerators, ISABELLE will be operated from a single control center. The control system will be distributed and will incorporate a local computer network. An overview of the conceptual design of the ISABELLE control system will be presented.

Keywords. Centralized Plant Control, Computer Control, Process Control, Microprocessors, Minicomputers, Particle Accelerators.

I. GOALS OF THE ISABELLE CONTROL SYSTEM

ISABELLE is a proton accelerator/storage ring. From the point of view of control system designer, it is large (3.8 km circumference) and complex (approximately 30 000 monitor and control variables). ISABELLE's function is to provide proton-proton collisions for the use of elementary particle physics experiments. In other words, ISABELLE is a utility. Thus the goals of ISABELLE's control system are more akin to those of the control system of a public utility or pertrochemical plant than they are to the goals of the data acquisition and control systems of the avant garde experiments ISABELLE serves.

ISABELLE comprises a number of systems. There are approximately 1000 superconducting magnets cooled by a large cryogenic refrigeration system and powered by over 300 power supplies. Beams of up to 8 amperes of protons in each of two rings circulate in a beam pipe evacuated by a large distributed vacuum system. There is an injection system which transports protons from the present Alternating Gradient Synchrotron into ISABELLE. A stacking rf system is used in the accumulation of beam current and an accelerating rf system accelerates the protons up to 400 GeV. The beams are stored for periods of one to two days, providing proton-proton collisions during that entire period. There is an extraction system which ejects the remaining protons at the end of the useful beam storage time, and transports them to a beam dump. The control of these various systems is unified by the ISABELLE control system - a local network of distributed minicomputers communicating with devices via a process data highway.

The goals of the ISABELLE control system can be simply stated:

1. All ISABELLE systems will be operated from a single control center.

2. In the case of the loss of all or part of the control computer network, accelerator devices will continue to operate at the most recent setpoint.

3. The hardware and software interface of devices to the ISABELLE control system will be implemented in standard ways.

4. The generation of programs and the testing of new devices should be routinely done by control system users.

5. The control system should be capable of phenomenologically modelling the state of ISABELLE - both in a predictive and historical sense.

A. Discussion of the Control System Goals

1. All ISABELLE systems will be operated from a single control center. This is by now standard practice in accelerator control systems. The existence of local control rooms, dedicated to a single system, is not incompatible with this goal. However, any computer control capability available in a local control room should also be present in the ISABELLE control center.

2. In the case of the loss of all or part of the control computer network, accelerator systems will continue to operate at the most recent setpoint. In addition, such a loss will not, of itself, cause the beam to be aborted. This principle or goal has important implications for the control system design. A

---

*Work performed under the auspices of the U.S. Department of Energy.

necessary condition for its achievement is that no feedback loop be closed through the control computer network. Such a goal has only becomes feasible because of the dramatic drop in the price to performance ratio of digital electronics. DDC (direct digital control) loops can be implemented in the devices which need them. The control system need no longer be involved in softwear servo loops. In the classic control engineering terminology, ISABELLE's control system is an open loop system.

In actual practice, this goal will be ignored from time to time. The feasibility of digital feedback loops can certainly be tested by using the control computer network. Once the feasibility has been proven however, such loops will be implemented in dedicated hardware.

3. The hardware and software interface of devices to the ISABELLE control system will be implemented in standard ways.

The ISABELLE control system can itself be considered a utility. It supplies a digital communications path and common computer facilities. Devices connected to the control system must satisfy appropriate connection standards for both hardware and software. The importance of protocols and interface standards is well known in the distributed data processing field. This is conceptually a similar problem. Consider the connection of a new type of magnet power supply to the control system. The power supply must have the appropriate plug through which the appropriate pulse trains are transmitted to communicate with the appropriate softward driver and data base module in the local computer node. The same communications hardware is used by many devices and device designers, hence the need for hardware standards. Once a device is incorporated in the control system, the device associated software such as device drivers and data base modules can be used by many programs and programmers, hence the need for software standards.

4. The generation of processes and the testing of new devices should be routinely done by the control system users.

It has always been true that one of the persons who is more knowledgeable than most about the capability of a device is the device designer himself. The SPS took advantage of this truism by implementing an easily learned computer language (NODAL) which allowed the device designer to become the applications programmer who used the device. The developments that have taken place since the SPS control system was designed have increased both the need and the opportunity for this amalgamation. The need has increased because the use of microprocessors and other LSI (VLSI) techniques has increased the sophistication and capabilities of the devices themselves. The opportunity has increased because the use of imbedded microprocessors has increased the level of sophistication of the software abilities of the device designers.

In most cases, the testing of new devices cannot be done during production running, when ISABELLE is producing luminosity for the use of elementary particle physics experiments. Thus the testing of devices in the realistic accelerator operations environment will occur at infrequent times and for restricated periods. To maximize the effectiveness and efficiency of the use of these periods, the device designer is the appropriate resource of which to take advantage. He must have at his hand comprehensive debugging and monitoring facilities. It is worthwhile noting the common experience that interpretive computer languages (BASIC, NODAL) are very useful in the testing and wringing out phase of device construction.

5. The control system should be capable of phenomenologically modeling the state of ISABELLE - both in a predictive and a historical sense.

Experience with SPEAR has shown the accelerator community that it is possible to control the global parameters of an accelerator, rather than the individual specific control elements. Such operation is only possible through the use of a mathematical model of the accelerator which exists in a computer of the control system. This model transforms the changes in accelerator parameters into increments of currents in power supplies, phase and amplitude of the rf accelerating system, etc. If the changes are so large that they can only be accomplished by a series of incremental changes, then the model calculates the increments, and then implements the series under open loop software control. Such global control is clearly a useful feature. After all, it is an accelerator that is being controlled, and it stands to reason that the language of the accelerator (beam energy, beta function, crossing angle, chromaticity, etc.) should be used, rather than some more arcane engineering language (volts, amperes, gauss, etc.).

For ISABELLE, this feature is more than useful, it is necessary. ISABELLE is complex and fragile. An operator needs to know the consequences of actions before the actions are implemented. If a beam manipulation will cause the beam to exceed the dump profile, the operator has to be informed of that fact before the manipulation is performed. Such proposed actions can be tested in the accelerator model, and the implications can thus be made known to the operator. As the predictive accelerator model is refined by operations experience, it will become an indispensible element in the operation of ISABELLE.

The historical model of ISABELLE is also a critical necessity. It will happen that the accelerator will be running and all at once, an alarm will be actuated and there will be no beam in the machine - because the beam extraction system will have been triggered. What has happened? Certainly the option to go back and do it again must only be used as a last resort - especially if quenching superconducting magnets is involved. It must be possible to phenomenologically reconstruct the state of ISABELLE and analyze what was the sequence of events which lead to the beam extraction. There are two requirements for this reconstruction process. The first requirement is data. Since ISABELLE is so complex (See Table 1.), and since many variables can change rapidly, the amount of data associated with such an event might be quite large - of the order of 5 megabytes, for example. In addition, since the occurrence of such events is by definition unpredictable, the data acquisition facility must be constantly operating. The second requirement for the reconstruction process is a model/procedure to analyze the data associated with such an event; i.e., the historical ISABELLE model.

II. INTRODUCTION

The basic design concept of the ISABELLE Control System is that it will be a distributed system. This conclusion is based on ISABELLE's size and complexity. The rationale for this conclusion has been well put by the authors of a design report which came to a similar conclusion in the case of the control system for a large steel manufacturing plant:

"Automatic control of the modern steel mill, whether achieved by a computer based system or by conventional means, involves an extensive system for the automatic monitoring of a large number of different variables operating under a very wide range of process dynamics. It requires the development of a large number of quite complex, usually nonlinear, relationships for the translation of the plan variable values into the required control corrections commands. Finally, these control corrections must be transmitted to another very large set of widely scattered actuation mechanisms of various types which, because of the nature of the steel manufacturing processes, involve the direction of the expenditure of very large amounts of energy. Also, plant personnel, both operating and management, must be kept aware of the current status of the plant and of each of its processes.

Such a [distributed] system should have many benefits that can be of great value when installing, operating, or altering the system. Some of these benefits are:

1. Flexible system configuration - distributed subsystems may be modified, replaced, or deleted without upsetting the rest of the system.

2. Graceful degradation - failure in one or more components or subsystems will not cause the entire system to fail.

3. High systems reliability due to:

Table I. There is a total of 21184 monitor variables and 10364 control variables in ISABELLE.

Device	System	Number
Monitor Variables		
Coil Voltage Dipole	Magnet	2928
Coil Voltage Quad	Magnet	2712
Magnet	Magnet	300
Orbit Position	Beam Monitoring	444
Lead Voltage Drop	Cryogenics	1024
Lead Temperature	Cryogenics	612
Pressure Transducers	Cryogenics	72
Flowmeter	Cryogenics	48
Magnet Temperature	Cryogenics	456
Gauges Ins. Vac.	Vacuum	324
Gauges UHV	Vacuum	1416
Ion Current Pump	Vacuum	1416
Bakeout Temperature	Vacuum	2832
Sector Valves	Vacuum	84
Turbopumps	Vacuum	168
Clear Electrodes	Vacuum	1416
Turbopumps Ins. Vac.	Vacuum	132
Gate Valve Ins. Vac.	Vacuum	648
Radiation Monitor	Access	162
TI Pump Current	Vacuum	1416
Instrumentation	Beam Monitoring	24
Temperature Sensors	Cryogenics	150
Pressure Transducer	Cryogenics	150
Flow Heater	Cryogenics	12
Vibration Analyzer	Cryogenics	36
Impurity Analyzer	Cryogenics	2
Digital IO	Cryogenics	500
Transfer Syst. Total	Injection	1000
Ejection Syst. Total	Ejection	500
Cavities	Radio-frequency	40
High Level rf	Radio-frequency	120
Low Level rf	Radio-frequency	40
Control Variables		
Power Supply 300 A	Power Supplies	144
Power Supply 100 A	Power Supplies	96
Power Supply 50 A	Power Supplies	252
Lead Flow	Cryogenics	612
Magnet Flow	Cryogenics	12
Lead Voltage	Cryogenics	1024
Valve Control	Cryogenics	48
Sector Valves	Vacuum	84
Ion Pumps On/Off	Vacuum	1416
Ti-Pumps	Vacuum	1416
Bakeout Heater	Vacuum	2832
Turbopumps UHV	Vacuum	84
UHV Gauges On/Off	Vacuum	1416
Turbopumps Ins. Vac.	Vacuum	66
Gate Valves Ins. Vac.	Vacuum	324
Power Supply 5 kA	Power Supplies	2
Power Supply 300 A	Power Supplies	10
Power Supply 50 A	Power Supplies	16
PID Control Loops	Cryogenics	60
Digital Control	Cryogenics	200
High Level Analogue	Radio-frequency	70
Low Level Analogue	Radio-frequency	80
High Level On/Off	Radio-frequency	60
Low Level On/Off	Radio-frequency	40

a. Easy to add parallel redundant units and subsystems which can be incorporated to back up and duplicate the functions of the main components and subsystems.

b. Transmission of partially processed plant information allowing:
   (i) Decreased data rates since processors are distributed to functional areas and only processed information need be sent between any two subsystems rather than raw data as formerly.
   (ii) Use of error detection codes which allow any fault or casualty condition in the system to be detected and identified by the processor in its area of responsibility.
4. Lower cost due to:
   a. Simplified hardware configuration packaging since processors need not be large due to the reduced processing requirements of each processor.

   b. Simplified software because functions are carried out by several small, locally responsible processors, not by a large machine that must perform all of the control functions and calculations within the entire control system.

   c. Large scale integration technology.

   d. Multiple use of standard components. Many different subsystems can use identical hardware to perform varied functions.

   e. Ease of incrementally increasing capability since units may be added to the system without drastically interfering with the functions of the rest of the system.

   f. Simplified installation since common data channels can be used for processor to processor communication. This eliminates the need for individual multiple-wire cables between any two units."
   (Purdue University Project Staff 1977).

What are the various elements of the distributed system used to control ISABELLE? To answer that question from a system viewpoint, consider the most common task that the control system will be called upon to accomplish, namely to allow an operator to control or monitor a device. (See Fig. 1.). In accomplishing this task, the operator normally interacts with the system at a control console, from which he can cause a program or command to be executed. Information concerning this command is communicated from the control console computer to a local computer over the computer network communication system. Resident in the local computer is a piece of software called a data module. The command sent to the local computer informs the data module what the operator wants. The data module communicates over the process data highway to the device. The command is executed by the device, and information of interest may be passed back through the system to the control console and communicated to the operator. All these communications and their corresponding protocols and interfaces are transparent to the operator. The operator thinks he is communicating directly with the device.

Thus, the control system can be conceptualized as a control center (control console/computer), a computer network, and a process (or device) interface system. Each of these three elements will be considered in more detail in the following pages.

III. CONTROL CENTER

The Control Center is where operators interact with ISABELLE qua system. The Control Center is composed of a number of elements:

1. Control Consoles.

2. Data base Facility and Program Library.

3. Analytic Computation Facility.

4. Program Development Facility.

5. Alarms and Access.

1. Control Consoles: In the language of the control engineer, a control console is known as the man/machine interface. The International Purdue Workshop on Industrial Computer Systems postulates two subsystems that make up a control system. They are the personnel subsystem (operator, management) and the machine subsystem (computers, controllers, industrial plant or process, etc.). "The Man/Machine Interface (MMIF) is that boundary between the two subsystems across which information and control manipulation flows. In physical terms, it is typically the face of a console which contains displays and keyboards of various types. The hardware and software behind this boundary participates in translating human inputs to control signals and process inputs into data displays for humans." (MMIF Committe 1978).

At ISABELLE, the control console (MMIF) will be a computer. Accelerator experience shows that there are two general kinds of accelerator operation-development and production. In development running, the operator is trying to push the state of the art [or, as (M. Hine 1979) points out, to repair the broken accelerator]. Change is the operator's goal, and also, to some extent, his technique. In production running, the operator is trying to maintain stability of operation. This operator's goal is integrated luminosity for elementary particle physics experiments. The existence of these two kinds of operations implies two kinds of operators and two sets of possibly overlapping control console requirements.

Although not strictly within the area of interest of this paper, the control center of ISABELLE will contain more than just the computer control system. Operators will need information from the beam monitoring and rf accelerating systems to run ISABELLE. Much of this information is of analogue nature, or is obtained from information processing equipment too expensive or bulky to duplicate at every control console. Experience with other accelerators indicates that it would be quite useful to have the low level rf system, the beam monitoring information processing equipment, and the control consoles in a single area or in contiguous rooms. The ISABELLE control room will be located close to the ring and cable runs between the control them and this instrumentation (pickup electrodes for longitudinal and transverse Schottky scans, beam current monitors, etc.) will be kept short.

The ISR Experience indicates that one operator (and therefore one console) should be dedicated to each of the storage rings. In addition the Engineer-in-Charge (EIC) or head operator should have a console. During production running, the EIC is charged with overseeing the operation of the colliding beam facility. The charge of the other two operators is clearly to keep the blue and yellow[1] rings stable. The SPS experience indicates that an additional console is needed for backup, and for the maintenance and development of new hardware and software for operations. "While a considerable amount of the program writing and modification can be carried out on the terminals of [a time-shared program development system], a console is necessary for testing and de-bugging if displays are involved, which is the case for most applications programs." (Crowly-Milling 1978). Thus, the conclusion is reached that four consoles are necessary.

Details of the control console design are not yet fixed. Following the example of the SPS, each control console will include a computer system. Operators will call accelerator application programs into execution which will, in turn, communicate with the various area computers and utilities for the purpose of information and control. Since the operator is interacting at the console with the accelerator, the applications programs can be quite large and complicated. Thus, the limitations of the addressing space set by sixteen bit memory may be too confining. Thirty-two bit computers are an obvious solution, and may very well be a cost effective solution.

Color video terminals are cost effective devices for allowing an operator to obtain an overview of the status of complicated systems. One can take advantage of the increased information per unit area available due to color (Crowley-Milling, 1978). This utility is available in a number of ways. Lists of numerical values can be color coded to show actual, set, transition, and out of limits values. System flow diagrams can be color coded with respect to both process flow and status. Of course, one can simply put more curves on the same "sheet" if the curves are drawn in different colors. This latter utility has only recently begun to be truly a cost effec-

___
1. Blue (yellow) is the designation for the clockwise (counterclockwise) ring of ISABELLE.

tive possibility. One of the problems with color graphics has been the lack of a good, fast hard copy facility. Such units are now appearing in the market place, although they are still expensive. However, one can conceive of a single unit in the ISABELLE control room which is shared among the control consoles.

The control console includes a computer system. A terminal to communicate with that computer system is necessary. The combination of CRT and keyboard satisfies that need. Cursor control via a track ball has become almost ubiquitous in accelerator control console design. ISABELLE will follow that lead. Computer controlled touch panels allow one to have as many on-off "buttons" as needed and still use very little console space. One can call programs into execution and turn on and devices via the touch panel. The former utility appears to be one of the most used features of the SPS control consoles.

The use of a light pen might be considered. One of the common tasks an operator has during development running is to modify a time based function such as is used in a function generator controlling a power supply driving a ramp, for example. It is hard to do this using a track ball - drawing a two dimensional curve with a track ball calls for some dexterity. It would be easier to "draw" the curve by hand on the "face" of a crt. A light pen allows just such an operation.

Most of the above-mentioned features are needed in both types of accelerator operation. This is because the MMIF as interface to the control system has been emphasised more than the MMIF as interface to the accelerator. Thus, much of the above mentioned elements of the control consoles are conceived of as computer peripherals. When we begin to consider some of the higher bandwidth information, the differences between production and development running begin to appear. These differences are due, in part, to the differences in the type of operator who is controlling the accelerator during the alternative types of running. The development operator is the machine scientist (or rf engineer, cryogenics system designer, etc.). In trying to push the state of the art, he is trying to manipulate parameters in untried ways or combination. Thus, by hypothesis, well tried and de-bugged software is not available. Since the techniques being attempted are new, ways of presenting the data are also not available. Combinations of parameters imply a need for multiple graphics presentations. In addition, since the accelerator physicist is manipulating beam, he will need more analogue readout devices, such as fast oscilloscopes, than the production operator. These needs of the development operator imply that at least one of the consoles should have more elaborate fast analogue presentations than the others. In addition, there should be a language available to the console user for easily writing programs which has de-bugging and graphics facilities. The interpretive language NODAL, used at the SPS, appears to be an obvious candidate to satisfy this particular need. BASIC may also be considered.

2. Data Base Facility and Program Library

In the largest sense of the term, a data base facility encompasses the program library. However, it appears to be common useage among accelerator users that data base refers to the parameters that describe the accelerator and the program library is the repository of programs used in operating the accelerator.

Any data base philosophy implemented in a distributed system is a compromise between two mutually exclusive needs-the need for a centralized data base and the need for a distributed data base. The advantage of a central data base is that the response to monitoring interrogations is quite fast. However the information obtained may be old. Maintaining a central data base in a distributed system necessitates a great deal of communication, which in turn puts high speed requirements on the computer communication technique. If one wants to take a "snapshot" of the state of the machine, it is only necessary to make a archival copy of the central data base or some portion thereof. This can occur quickly because there is no need to communicate to many distributed devices. The problem of software surveilleance is much simplified, given a central data base. Surveillance programs can monitor the data base, rather than the devices. A major disadvantage of the central data base is that any parameter change must be communicated to it. Thus, parameter changes at remote locations must communicate with the central data base.

The SPS has implemented a distributed data base in its use of "data modules." A data module is a piece of software resident in the computer system closest to the actual physical device. All information concerning the device (addresses, operations, limits, set points, etc.) is contained in the data module. The set of all data modules thus contains the accelerator data base. The advantage of this approach is that the information obtained from the data base in "immediate." Parameter changes can be effected without needing to be communicated to a central facility. The disadvantages are that "snapshots" take much longer to do and surveillance puts more burden on the communications system.

The advantage of being able to effect a parameter change without having to communicate with a central facility appears to be the most weighty factor. A strong point in its favor is the fact that all the information concerning a device is located in one piece of software. This clearly makes the device much easier to de-bug in both the hardware and software aspects. A review of the goals of the ISABELLE control system shows that goals 2, 3, and 4 tend to put their weight on the side of the distributed data base approach.

Thus, the conclusion is reached that ISABELLE will follow the SPS example and use a distributed data base.

Can some of the positive features of the central data base be retained? Yes, if it is decided to take advantage of ISABELLE's stability of operation. There can be two data bases in the ISABELLE computer system. The primary working copy will be the distributed data base in the distributed data modules. A secondary copy can be a central data base, resident in the "modelling computer." It will be necessary to update this secondary data base, which will mean heavier traffic on the computer network communications, which could conflict with the usage of the communications system for higher priority traffic. One could resolve this priority problem by in fact giving a lower priority in the message transfer system to messages which have to do with updating the central data base. One might incorporate this data base updating traffic into the message transfer self surveillance traffic (Crowley-Milling 1978). There is still the problem with the "aging" of the central data base, but ISABELLE must be stable for normal operation. Thus the fact that the data is the central data base may be "old" should not be critical. The immediate value of a parameter is always available in the data module in any case.

Should the program library be centralized or distributed? As far as data modules are concerned, an archival copy in a central location would be useful, but the backup copy should be located on disk or some kind of permanent media at computer where the data module is used. The difference between "backup" and "archival" is mostly the difference in the age of the data in the data module, the backup copy containing the more recent information. The actual programs themselves must be identical.

Should the applications program library be centralized or distributed? In general, applications programs are to be run in control console computers or other computer systems in the control center. These computer systems can be quite large, and disks or other storage media will probably be a part of each system. Thus, a distributed program library would appear to be the obvious choice. However, most operators will interact with the control consoles. If each console has its own program library, then the philosophy of having all consoles appear identical to the computer system is difficult to maintain. An operator may make a change to a program on his disk. This change would then have to be implemented on the disks of the other consoles. Unless such techniques can be implemented easily, the uniformity of the control consoles will soon be lost. A simpler approach would be to have a single program library (Crowley-Milling, 1975).

3. Analytic Computation Facility - The Modelling Computer

This facility specifically addresses the goal of being able to model ISABELLE in both the predictive and historical sense. Logically, it is a facility capable of performing complicated accelerator simulations in real time. Physically, it may be a combination of a link to some central scientific computing facility and a 32 bit (or larger) computer which is part of the ISABELLE control computer network. More complicated simulations could be done via the link at the central scientific facility, possibly in batch mode. The computer in the control computer network is a more dedicated facility. It would be used for doing simpler calculations (but not simple calculations - thus the need for 32 bits). Since the model for the accelerator exists here, this is a natural place to have the data base available - allowing convenient comparison between theory and reality. This feature is clearly of use for the purposes of surveillance also.

The phenomenological models of the accelerator will take some time and actual operational experience before they can become reliable aids in the running of ISABELLE. The impact of a reliable predictive model which would be interrogated whenever substansive changes are made must be investigated carefully. Clearly, a model which is interrogated whenever any change, no matter how small is made, possesses most of the disadvantages of the central data base facility, plus its own disadvantage of imposing some delay due to a necessary calculation or table lookup to see if the requested operation is acceptable.

4. Program Development Facility

Control consoles are too precious a commodity to be used as common terminals in the writing of programs and reports. In addition, many programs, such as surveillance programs, systems programs, data modules, etc., have little or no need of the graphics and other utilities of a control console. A large computer with a time shared operating system to support a number of terminals is an obvious candidate for a cost effective facility which will support a great deal of program development. In addition, it could act as a backup for the modelling and data base facility to increase the over all availability of these critical control system utilities.

Since most of the actual programs used in the ISABELLE control system will be developed on this facility, this is an appropriate place to discuss what languages will be used in the ISABELLE control system.

The field of computer programming languages is very rich, and getting richer. Philosophies of language use on computer systems run from assembly language only systems used in dedicated or embedded systems (such as device controllers) to the anarchy of large computing facilities with the capability of supporting almost any conceivable computer language. What should be ISABELLE's philosophy? The projected lifetime of ISABELLE and its control system is between two and three decades. The necessity that programs written in the first decade of ISABELLE's existance be understandable and modifiable requires that they be written in a language which will continue to be supported by the ISABELLE control system during ISABELLE's lifetime. In order that goal 4 be achieved, the languages chosen should be useful and popular at least among the ISABELLE control system users. Many of the design personnel of ISABELLE are acquainted with FORTRAN and BASIC. Both of these languages are standardized and well supported by the industry. The language standards which are or will be available include real-time extensions for process control and synchronization mechanisms for operation in a multi-task environment. Both these languages appear to satisfy the requirements of popularity, utility, and lifetime needed for ISABELLE. It can be argued that NODAL may be of greater utility than BASIC because of its greater graphics and string manipulation utilities. Certainly its development in an accelerator environment makes it a strong contender for replacing BASIC at ISABELLE.

In addition to the aforementioned languages, assembly language will of course be available to the knowledgeable user. However, the field of computer science has become very enamoured of procedural languages - specifically PASCAL-which afford more transportability, meaning that PASCAL is supported by different computer manufacturers, than assembly language and more utility for systems programming problems than FORTRAN, BASIC, or COBOL. If the present enthusiasm of the computer manufacturerers to support PASCAL continues, making a decision for or against its use may become moot. Even though it may not be used by control systems users, PASCAL seems an obvious choice as a more efficient alternative (because it is a high level language) to assembly language for systems programming problems.

Thus the languages proposed for use in the ISABELLE control system are FORTRAN, BASIC (NODAL?), PASCAL, and Assembly Language. In such an environment, two additional points must be made. An efficient, easily used word processing facility for software documentation must be available. Perhaps a managerial policy that no program will be allowed to reside on the operations disk for more than two weeks unless a documentation file is available might be enforced. The second point is that subroutines written in one language should be callable by another. For compiled languages, and for subroutines called by an interpreter, such facilities are already available. However BASIC and NODAL are interpretive, and facilities to call an interpreted BASIC subroutine from a FORTRAN, PASCAL, or Assembly language Program do not yet exist - to the author's knowledge. A simple solution is to have both an interpreter and a compiler for BASIC. Such facilities are available from at least one computer manufacturer. A compiler doesn't yet exist for NODAL, but could be written if deemed necessary.

5. Alarms and Access

As the design of the ISABELLE Control system becomes more developed it may be that the alarms and access functions become more separated. For now, it appears that the functions are complementary enough to be connected together in a single computer system.

The alarms function is intimately connected with the surveillance function. One surveys to find parameters which are out of limits, and then broadcasts an alarm. In the environment in which a central data base is available, much of the surveillance of accelerator parameters can occur via that data base. Thus a parameter surveillance program could run in the central data base facility, and alarm states can be communicated to the alarms computer. Some parameters will possess time constants such that surveillance in the central data base is deemed too slow. In those cases, surveillance can occur in the data module, or even in the device itself. In this latter case, an interrupt or service request facility must be available so that the device can gain the attention of the local computer to inform it of the alarm condition. The alternative to this interrupt facility is a polling scheme where the local computer polls all devices asking if they have an alarm condition. In any case, it is clear that alarms can come from various sources to the alarms computer. In addition to parameter surveillance of the accelerator, a similar function must be performed for the central computer network. The communications links must be surveyed. There are two candidates for this function. The obvious one is of course that computer which is dedicated to the communications system. Another possibility is the alarms computer because, like the communications or message handling computer, it has systems wide responsibilities and is naturally in communication with all computers of the network.

Having received notification of the alarm state, the alarms computer will then transfer the appropriate alarm message to each control console, which will have a video terminal dedicated to alarms. Depending upon the priority of the alarm, the operator may decide to follow different courses of action - which may range from mere acknowledgement to emergency beam abort. In addition to the alarms terminal at each control console, the alarms and access computer will have its own console in the control room. Alarms messages will also be recorded on this terminal also.

During beam off periods, access in ISABELLE will be controlled from the access and alarms console in the main control room. The access console will have video monitors which can be switched to any of the TV cameras covering the various points of personnel entrance and exit. The operator can thereby monitor access.

An important aspect of the access system is that it is most used during maintenance periods. However, accelerator maintenance includes control system maintenance, which may imply downtime for the network communications system. Thus the access system needs to have communications to the central access console which are independent of the control computer network. (The alternatives of a guard at each entrance/exit or uncontrolled entrance/exit are not considered viable for economic or operational reasons.)

IV. CONTROL COMPUTER NETWORK

The distributed systems approach upon which the ISABELLE Control System design is based uses the concept of a widely distributed set of functions carried out by small and relatively inexpensive computer systems (Purdue University, 1977; Crowley-Milling, 1975; Dimmler, 1978). These distributed parts are unified into a system by means of the control computer network which supplies the means of communication, and the various software and hardware utilities to support parallel processing and shared resources. A computer network in an environment such as ISABELLE is considered to be a local network. A local network is local in two senses:
1. It is generally owned by a single organization.

2. It is geographically local; i.e. distances are on the order of a few miles (Thurber, 1979).
Both of these attributes of a local network have positive aspects that reduce the complexity and, therefore, the cost of the network implementation - especially in ISABELLE's case. Since ISABELLE is a new single organization, a single type of computer hardware/software can be managerially imposed upon the solution. Thus the implementation of specific protocols and interfaces can be kept to a minimum. The presence of different operating systems or instruction sets could necessitate multiple software interface implementations, with a concomitant increase in the cost of implementation and maintenance. Geographic locality can mean that the error rates and delay of communication lines are drastically improved over those used for non-local distances. Thus specifications on delay and throughout can be met with less expensive technology than would be needed for similar performance in non-local networks.

6. Computers

Figure 2 shows a block diagram of the control computer network with the functions ascribed to the different nodes (computer systems) of the network. The architecture shown is referred to as a star, where the nodes at the points of the star communicate through a central communication system.

That portion of the nodes which are part of the control center have already been discussed. The assignment of the other twelve nodes has been based on the philosophy that there are functions of sufficient complexity and isolation that a computer system can be effective in the autonomous control of that function. Thus, for example, the cryogenic system or vacuum system can be operated independent of the power supply system, for the most part. Some of the functional assignments of the different nodes are obvious - rf, cryogenics, power supplies, and vacuum. The separation of injection from ejection is less obvious, but is made rational on the basis of the criticality of the ejection function from the point of view of personnel and accelerator safty. Although the computer system itself will not be involved in the triggering of the extraction sequence, the monitoring of that sequence before and during the event is of vital importance for the historical phenomenological model. In addition, the injection computer assignment is still under discussion. It may be that that computer is the AGS control computer itself (presently a PDP10). Such a design decision clearly subtracts from one of the advantages of a local computer network. A possible solution is to replace the PDP10, but this may not be cost effective. Another solution is to use the concept of the gateway (or interface) computer. This is a computer (of the same type used in the rest of the ISABELLE network) that would be placed between the ISABELLE computer network and the AGS PDP10. Thus, as far as the ISABELLE network is concerned, all the computers are identical. The gateway computer then performs the necessary translations between the AGS control system and the ISABELLE control system. While such a solution does not abrogate the basic problem of needing two software interfaces, it does localize the translation function into a single computer system which is independent of the AGS and ISA control functions. Such a gateway computer is clearly a minimal cost hardware configuration.

The assignment of a computer system to each sextant bears more investigation, and is intimately connected with the design philosophy of the process interface communications topology. For example, does the ISABELLE vacuum computer communicate with vacuum devices over its own communication link? Or does it communicate with the appropriate sextant computer which then communicates with the vacuum device via a link shared by all ISABELLE subsystems? In this latter case, operation of the vacuum system, qua system, is not possible unless the network communication system is functioning. This fact then impacts the availability specification of the network communications. Experience with the SPS shows that the fraction of accelerator downtime attributable to the control system can be held to the 1% or less level (Crowley-Milling, 1978).

What kind of computers should be used? Considering the present marketplace, this question reduces to the question 16 or 32 bits? The answer is 32 bits because the complexity of ISABELLE requires large application programs which need the addressing capabilities of the 32 bit machines. While it is true that large programs can be run on 16 bit machines using overlay techniques, the efficient use of such techniques requires more sophistication on the part of the user programmer than is consistent with goal number 4. It is possible that a mixture of 16 bit and 32 bit machines may be used. One computer vendor makes a 32 bit machine which, in addition to its own instruction set, executes the instruction set of a 16 bit computer manufactured by the same company. This would allow ISABELLE to assign 32 bit machines to functions where the large address space is needed and 16 bit machines where it is not. On the other hand, the 32 bit marketplace is an active one, and a cost effective solution of using 32 bit machines at all nodes of the network may very well be possible.

7. Network Techniques

What kind of local computer network (LCN) techniques will support the star-like configuration shown in Fig. 2.? Two techniques appear feasible - the message handling computer and the contention system. The SPS control system uses a message handling computer. Examples of contention system are Ethernet and Hyperchannel.

The message handling computer that the SPS uses is a polling, store and forward technique. Each node on the star is polled in succession to see if it has a message. If node A replies yes, the message is received in the central computer and retransmitted to the intended receiver (node B; see Fig. 3). Another possibility along similar lines is the circuit switch. In this case node A tells the message handling computer that it has a message for node B. The message handling computer then sets up a physical circuit between A and B over which the message is sent with no pause at the MHC. In an LCN environment, there is an important difference between these two. Because of the necessity to store and retransmit, the former technique becomes slower than the circuit switch for long messages. For short messages however, the store and forward technique may be faster - due to the necessary overhead to set up the circuit of the latter technique.

The contention technique is presently receiving a great deal of attention in the LCN community. (See Fig. 4). In this method, all nodes are connected to a common high bandwidth coaxial cable. A station wishing to transmit senses whether or not there is a carrier signal on the coaxial cable. If there is, a message is already on the line, and the station waits until the line is quiet. If (or when) the line is quiet, the station then transmits its message. All message include error checking words to detect a corrupted message - as would occur if two or more messages collide with one another. If collision occurs, the message is retransmitted. This technique possesses good characteristics for both short and long messages, as well as the feature that broadcasting (one sender, multiple receivers) is obviously included. Broadcasting is difficult to incorporate into either of the message handling computer techniques. Broadcasting is clearly a useful procedure for program synchronization.

In terms of performance objectives, there are two measures of the speed of a communications scheme - delay and throughput (McQuillan, 1978). Delay is usually measured in terms of average response time, throughput in terms of the peak traffic level supported. In a control environment, which must respond to real accelerator events, it is delay that is more important than throughput. The difference between the two concepts is commonly described by the difference between a satellite link (delay = 0.25 sec, throughput = 1500 kbs) and a common voice grade telephone circuit (delay = 0.01 sec, throughput = 2.4 kbs). For the former case, the time to send a 100 bits is 0.25 sec, and, for 1000000 bits, 0.92 sec. In the voice grade circuit case, the corresponding times are 0.05 seconds and 417 seconds.

In the case of ISABELLE, what are the time constants of most interest? In terms of response time, the fastest event that the control system might be expected to respond to is a quenching magnet, whose typical time constant is of the order of 100 milliseconds. A more significant time is the period magnet takes to loose 1% of its field - which is approximately 5-10 milliseconds. It could be of great utility to be able to transmit a message through the network before the magnetic field has significantly changed. This would be a short message of the highest priority, and would set the delay specification. (This message would be in addition to the triggering of emergency quench procedures, which would be handled in a system separate from the control computer network). Thus a delay of 5 ms for the sending of the highest priorty message from one computer node to another would be set by this analysis.

What is a reasonable specification on throughput which would be set by accelerator operation? The SPS has done an in depth analysis of traffic in their computer network. The main traffic is among computers which, in the terms of this paper, are in the control center. In other words, traffic to and from the accelerator is not a major component of the total network traffic. However, throughput is measured in terms of peak traffic level supported, not average. In the case of ISABELLE the peak traffic induced by accelerator operations would be during the transmission of data acquired during an extraction event. Let it be assumed that there are approximately 2000 circular buffers of 1000 bytes each which are constantly acquiring magnet coil data. This data is only read out after an extraction event has occured. There are thus 2000 monitor point x 1000 bytes x 8 bits/byte $\approx 20 \times 10^6$ bits of information. To read this event onto a disk in one minute is an upper limit to the length of time one would want to wait. 20×10^6 bits in 50 sec means a throughput of 400 kbs.

Measured results on the SPS message handling computer show a delay of 4 milliseconds and a throughput of 688 kbs (using a technology which will be 15 years old when ISABELLE turns on) (Crowley-Milling, 1978; Altaber, 1978). The ISABELLE requirements appear to be conservative, in terms of what the technology has already achieved. A facility for monitoring the performance of the network communication system must be available. Such a facility would include diagnostic and debugging tools. In the case of a system based on a message handling computer, such a facility may be present in the message handling computer itself. In the case of a contention system a computer dedicated to monitoring and diagnosis of communications traffic may be necessary to fulfill this requirement.

8. The Process Interface

The process interface is a mixture of hardware and software. Specifically, the process interface is the device dependent software in the local computer, the communications hardware and software which connects that computer to the process device, and it is the data and command structures in the device itself.

It is important to point out that a device is not just the control electronics which is connected to an equipment; a device is the equipment and its controls. Thus, device design is not necessarily the responsibility of the Controls Group. That responsibility resides in the group building the equipment.

It is the process (or device) interface that will be impacted most by the distributed approach of the ISABELLE Control System. The advent of microprocessors and other LSI (large scale integration) techniques allows the device designer the option of making intelligent instruments (or devices). It is this development that has allowed the possibility of achieving a control system whose goal is to do only set point control. All DDC (Direct Digital Control), e.g. digital feedback, can be off-loaded from the control computer network to the device. Control algorithms will be resident in the hardware and software of the actual device itself. The only information necessary to be transmitted to the devices will be high-level commands (off, on, standby, receive or transmit function table, etc.) and data.

9. Device Design Philosophy

Many of the devices of ISABELLE will be contained in the equipment areas of the ring. These areas (see Fig. 5.) are not accessible during operation. In order that ISABELLE operate reliably, equipment in these areas must be designed to operate without human intervention. It is just in this kind of environment that the distributed approach can be effective - if the appropriate design techniques are followed. Devices should be self contained logically and not require the constant intervention of the computer system. Such intervention is possible, but it puts great demands on the throughput of the process data highway (the communications system between the computer network and the device). Thus, the keywords concerning device design in the distributed ISABELLE environment are self-containment and reliability. To achieve these goals, the philosophy of device design can be summarized as follows:

ISABELLE Device Design Philosophy

1. No device should require the local computer to be dedicated to a closed loop. If such a closed digital loop is necessary, it is the responsibility of the device designer to incorporate it in the device.

2. Only data and commands can be transmitted between the control computer and the device. Specifically, the device designer is not allowed the option of having programs down-line loaded into his device.

3. All devices will communicate with the computer network via a standard communication using standard protocols. Command and data structures used in devices will also conform to control system standards.

The last point, albeit specifically addressed to the device side of the process interface, also applies, in part, to the computer side. The data module is the combination of software driver subroutine and data base which contains all the information about the device known to the control computer network (limits, set points, commands, etc.). Any program seeking to control or monitor a device does so via the data module. Since this facility will be used by numerous individuals, it too must conform to control system standards with respect to data formats, command structure, and documentation.

10. Process Data Highway

The process data highway is the means of communication between the control computer network and the actual device or process. Previous accelerator implementations of the process data highway are Datacon (AGS), SEDAC (PETRA), Serial CAMAC (ISR), and the MPX (SPS). Of the four, only serial CAMAC has been used elsewhere. None of the other three can be considered more than single-site solutions to the more general problem of distributed process interfacing. Serial CAMAC has been used elsewhere, but suffers from the detriments of being expensive (Rausch, 1976), lacking in reliability (in the process envionment) due to its limited error checking capabilities and poor connector design, and being crate-oriented. This last detriment means that it is somewhat incompatible with the philosophy of having intelligence located in the controlled device. CAMAC itself is a mechanical crate and bus standard. Serial CAMAC is a means to communicate with a series of CAMAC crates. It is not a means to communicate with a series of autonomous intelligent devices.

As in the case of the local computer network techniques, no design decision has yet been made. A preliminary report on the subject is due in January of 1980. What are the functional requirements of the process data highway? A working group (WG6) of the Technical Committee on Industrial-Process Measurement and Control (TC65) of the International Electro-Technical Commission (IEC) has begun to circulate draft versions of the functional requirements of PROWAY - a process data highway for distributed process control systems. This is a rather extensive document, (British Electrotechnical Committee, 1979), too lengthy to reproduce here. Many of the requirements and definitions are applicable to the ISABELLE environment. This committee defines the optimum characteristics of a process data highway to be:

"1. Event driven communication which allows real time response to events.

2. Very high availability.

3. Very high data integrity.

4. Proper operation in the presence of electromagnetic interference and differences in earth potential and,

5. Dedicated intra-plant transmission lines."

Typical delays mentioned in the document are less than two milliseconds and throughputs mentioned are in the range of 30 to 1000 kbs. ISABELLE personnel are actively following the development of PROWAY.

What might be the architecture or topology of the process interface? As mentioned earlier, many devices in ISABELLE are contained in the equipment areas of the magnet enclosure. Other locations are the service building, equipment areas at each intersection region, the rf building, the cryogenic compressor building, etc. The process data highway must connect to the devices in these areas and then transmit information to a fewer number of locations where connection to the local computer will take place. An obvious topology would be to connect all the devices within a relatively short distance (~ 100 ft) to a data concentrator and then transmit over longer distances to a connection to the local computer (see Fig. 6). (In fact, this is the model used in the PROWAY functional specification). Such a technique decreases dramatically the number of wires (cables) that must be stretched over long distances.

Connection of the equipment alcoves to the local computer may occur in a number of ways, but two obvious candidates are shown in Figs. 7 and 8. The hierarchical star of Fig. 7 follows the example of the SPS configuration. All the equipment on a sextant connects to one of the sextant local computers. A problem with this configuration is that one cannot communicate with all vacuum or cryogenic devices unless the computer network communication system is functioning. However, due to the fact that each sextant can be communicated with in parallel with the others, the total throughput of information is six times that of a single computer collecting data from the entire ring.

Figure 8 shows a loop or broadcast connection in which a single dedicated system computer (cryogenics, vacuum) communicates over a single link to all the devices of that system. It is quite common that such links have redundant paths for reliability considerations. Such a topology solves the problem mentioned in the preceding paragraph, but at the expense of decreased throughput.

It may be that a combination of the two solutions is the best technical choice. The cryogenic and vacuum systems, because of their operation, require that they be functional for up to 2-3 weeks before accelerator operation. Systems such as beam instrumentation and power supplies are much more tightly coupled to the presence of beam when operation as an accelerator system is considered. The rf system is very local in any case (the low level rf system and power supplies are in the service building, the high level rf system is in the rf building). In addition, the beam instrumentation can have much higher throughput and lower delay requirements than the cryogenic and vacuum systems. Thus, the hierarchical star à la SPS may be the best technical choice for the beam instrumentation, power supplies, and magnet monitoring system.

What technique should be used to connect devices in an alcove to the concentrator? One possible candidate is IEEE-488. This instrumentation-oriented standard was designed to connect devices over relatively short distances to a computer. This design goal of IEEE-488 is very close to the kind of function needed in an alcove. In addition, the standard comes impressively documented and well supported by the LSI industry. Our experience with IEEE-488 indicates that it is much more compatible with microprocessor designs than CAMAC or DATACON. In addition, it has a limited interrupt facility (service request) which is necessary in an environment where the control computer network is not constantly monitoring a device. The connection between an intelligent device and interrupt capability is so strong that the IEC, in the glossary section of PROWAY document defines an intelligent station as one "which includes application units [devices] capable of initiating and controlling message transactions through a Data Highway." In fact, IEEE-488 interfaces will exist as part of the ISABELLE control system because of its acceptance by the instrumentation industry. However, a committment to use IEEE-488 as a process interface in addition to using it for various industry supplied instrumentation (digital voltmeters, sweep generators, network analyzers, etc.) must await a more detailed analysis of the cost effectiveness of the standard in the ISABELLE environment.

11. The Timing System

An accelerator is a single device comprising many systems, some of which must be synchronized or time ordered (Crowley-Milling, 1975). Specifically, during the acceleration period, the power supplies must be ramped in synchronism with one another and with the accelerating rf system. In addition, measurements of beam location from different orbit monitors must be synchronized, or at least time ordered. If a superconducting magnet reverts to the normal state, measurements of beam location, extraction, temperature rise, etc. must be time ordered. A clock is a necessity.

The rate of the clock lies somewhere between the rate at which protons circle ISABELLE (10 microseconds per revolution) and the characteristic time of a quenching magnet (~ 100 milliseconds). The present philosophy is to use a 10 kilohertz clock which is transmitted to each equipment area around the machine. Within the equipment areas, a local 10 kilohertz clock is locked to the master clock and transmitted over the ac power system using FSK (Frequency Shift Keying) techniques.

12. Review of System Goals

How does the system proposed meet the goals outlined at the beginning of this paper?

1. "All ISABELLE system will be operated from a single control center." A single control center for ISABELLE will exist. All systems connected to the control computer network will be operable from that control center. The existence of a central backup data base, program development and modelling facilities and extensive control consoles will make operation from that center efficient and attractive.

2. "In the case of the loss of all or part of the control computer network, accelerator systems will continue to operate at the most recent setpoint." The use of intelligent, self-contained devices and the policy that

the control computer network will not engage in direct digital control gives the accelerator an effective buffer in the case of a network failure.

3. "The hardware and software interface of devices to the ISABELLE control system should be implemented in standard ways." The enforcement of software and hardware standards and the checks and balances available by forcing all accesses to devices to go through the common data module is a direct and effective response to this goal.

4. "The generation of programs and the testing of new devices, should be routinely done by control system users." The use of common well known languages such as FORTRAN and BASIC will make programming very approachable by control system users. The use of computers with large addressing space will not make it necessary for the users to understand the complexities of overlaying and other esoterica for the implementation of large programs. The use of standard data modules, synchronization and data base procedures will allow the users a uniform approach to device control.

5. "The control system should be capable of phenomenologically modelling the state of ISABELLE - both in a predictive and historical sense." A powerful modelling facility exits for this specific goal.

V. ACKNOWLEDGEMENTS

Readers familar with accelerator control systems will recognize the debt owed to M. C. Crowley-Milling and his colleagues of the SPS. Colleagues associated with accelerators at CERN, FNAL, SLAC, and LASL have also contributed. The International Purdue Workshop on Industrial Computer Systems has been a fruitful contact with colleagues in the industrial community. Naturally, colleagues at BNL have contributed ideas, clarifications, and much appreciated constructive criticism. Too numerous to name, they are associated with the Accelerator, Physics, and Applied Math Departments, the Instrumentation Division, and, of course, the ISABELLE Project. Robert Frankel of the ISABELLE Project has been an invaluable resource in the development of the concepts described in this paper.

VI. REFERENCES

British Electrotechnical Committee (1979). Technical Committee No. 65, Intern. Electrotechnical Commission, Draft - Process Data Highway (PROWAY) for Distributed Process Control Systems, IEC 65A (Secretariat), London, Great Britain.

Crowley-Milling, M. C., (1975). The Design of the Control System for the SPS, CERN Report, CERN 75-20, p. 9.

Crowley-Milling, M. C., (1978). Experience with the Control System for the System for the SPS, CERN Report, CERN 78-09, p. 17.

Crowley-Milling, M. C. (1978). Experience with the Control System for the SPS, pp. 24-26.

Crowley-Milling, M. C. (1978). op. cit. p. 5; Altaber, J., Jeanneret, J-P, (1978). Measurements on the Control Network of the SPS, CERN Rept. CERN/SPS/ACC/78-1, Translated by F. Beck, CERN, Geneva, Switzerland (1978).

Crowley-Milling, M. C., (1978). op. cit. pp. 5-6.

Dimmler, D. G. (1975). ICAM - The ISABELLE Control and Monitoring System, Global Overview, BNL Report, CRISP 75-5; Lamport, L. (1978). Time, Clocks, and the Ordering of Events in a Distributed System, Communications ACM, $\underline{21}$, pp. 558-565.

Dimmler, D. G., (1978). Computer Networks in Future Accelerator Control Systems, IEEE Trans. Nucl. Sci. $\underline{NS-25}$, No. 2, pp. 974-988.

Hine, M., (1979). Private communication.

MMIF Committee, (1978). Guidelines for the Design of Man/Machine Interfaces for Process Control, Intern. Purdue Workshop on Industrial Control Systems, Purdue University, 2nd Rev. p. vii cf. "and what is the use of a book" thought Alice, "without pictures or conversations?" Lewis Carrol, Alice in Wonderland (1865) quoted by F. Beck, The Design and Construction of a Control Center for the CERN SPS Accelerator, CERN Report. SPS-CO/76-1 (1976).

Purdue University, Project Staff, (1977). Tasks and Functional Specifications of the Steel Plant Heierarchy Control System, Report #98, Purdue Laboratory for Applied Industrial Control, pp. 2-3 to 2-5.

Rausch, B., (1976). Cost Comparison Between Serial CAMAC and SPS-MPX System for Control Applications, CERN Rept. SPS-CO/Int/CC/76-6.

Thurber, K. J., Freeman, H. A. (1979). A Bibliography of Local Computer Network Architecture, Sperry Univac, St. Paul, Minnesota.

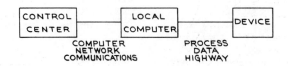

Fig. 1. Conceptual Computer Control System.

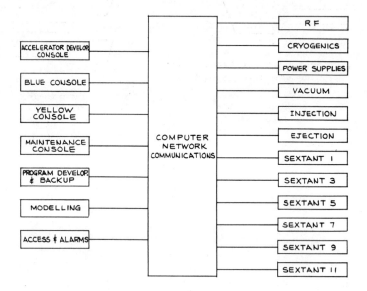

Fig. 2. Computer System Layout.

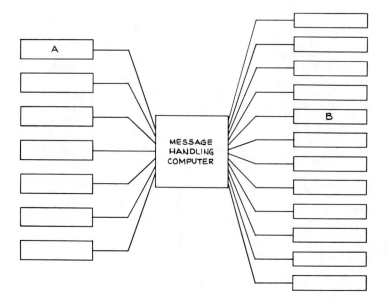

Fig. 3. Local Computer Network Based on a Message Handling Computer.

Fig. 4. Local Computer Network Based on a Contention System.

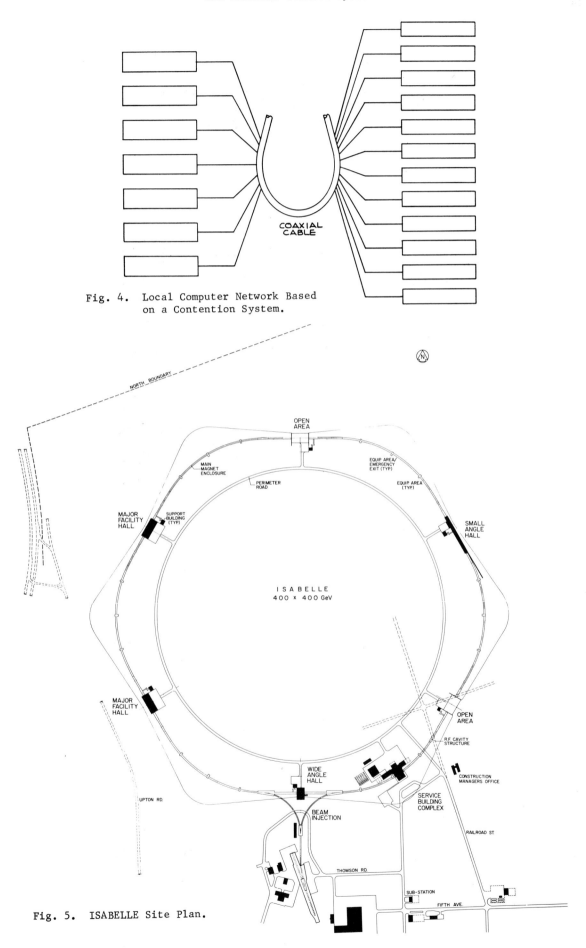

Fig. 5. ISABELLE Site Plan.

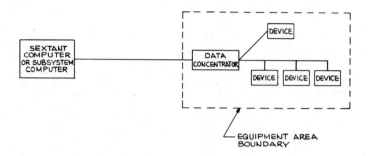

Fig. 6. Conceptual Diagram of Communication between a Local Computer and Devices in an Equipment Area.

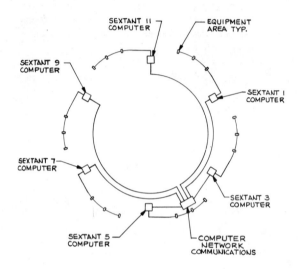

Fig. 7. Star topology for the Process Data Highway.

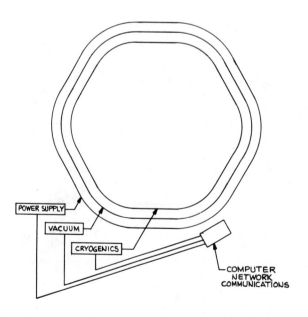

Fig. 8. Loop topology for the Process Data Highway.

DISCUSSION

Garcia: You mentioned the data base management considerations in this particular application. Could you expand on the subject? I think that this is probably one of the key items in a project of this nature, where I see that the control center should be able to query the Computer Network's Communication, at a functional level, on the state of the accelerator in a device independent manner and some low level device down at the other end should be able to do the adequate translation. It appears to me that, in a situation like this, there is room for application of many concepts of data base management that have existed for several years in the commercial field, except that you have to introduce the tremendously demanding consideration of real time applications. This is an area about which there is very little known and I would like to know what considerations you gave to this.

Humphrey: In fact, in the accelerator community we have been facing this problem in the past and it is the reason that the SPS (Super Proton Synchrotron) system is such a big breakthrough. The SPS has a distributed data base where the data base is contained in the device drivers. Other accelerator systems, specifically the AGS control system at Brookhaven and the Fermi National Accelerator, have a more centralized data base. The future of that course is that when you ask the data base for something, you get it right away; but now, if you actually try to control the device, you clearly have to talk to this data base, which is centralized, and then talk to the device which is distributed and there are delays due to communications and data base access.

We have at the AGS a process data highway that is capable, in terms of its hardware speeds and actually its software driver speed, of supporting communications in the three to 10 kilohertz region. But, when you went through the data base, you were reduced to three readings a second, which was just a disaster! The reason for that was that it was a very nicely constructed data base and you didn't have to know where a device was; you didn't even have to know many of its attributes. From the point of the user, it was very, very nice. From the point of view of real time control, it was a disaster. So I agree that we are now reacting against that in the accelerator community, or at least in the case of a couple of the accelerator control systems by going to this distributed data base philosophy. But it is exactly the point which you brought up of the real time constraints that is driving us. We see no solution at this point, other than we are really trying to follow what's happening in the tremendous development that is going on in data base structures. This is a hard problem that is fundamental.

Willard: You came down very hard in your specifications associated with device design on the fact that it was not permissible to have programs downloaded into devices from elsewhere within the network. In view of the economy of scale frequently associated with having program development on a single large machine, why are you coming down so strongly against downline loading of programs into devices?

Humphrey: The reason is that many of the devices in ISABELLE, specifically the ones in the ring, cannot be reached by people. When the machine is running they are not accessible. So, as a result, we are trying to conceive of these things as electronics; as something that is rock solid and cannot move. The only way to change those programs is to go down and yank out the ROM and put in a new one; that is our concept. In the case of devices that are not so inaccessible, we are not that big of an operation and we are fairly friendly so that we can modify this position. But I think that in the case of the inaccessibility argument, it is quite strong for us; it is really quite strong!

Willard: So the key issue, then, is the reliability associated with the programs that are in the device and, if you could solve that through multiple paths to downline load it and have control over the times at which you downline load, then you would not be terribly opposed to downline loading.

Humphrey: The reliability of communication paths doesn't bother me much. I think we can build those today. What I need more now are mechanisms for the control of the software that goes down there. It has to be software that is guaranteed to work, and the mechanism of using a ROM is the way I solved that. It's a disaster if a piece of untried software causes ISABELLE to yank its beam out, not a disaster that is going to kill people, but it's going to cost a lot of downtime and we are very very worried about that.

Lambert: Why do you feel that putting the program in a ROM guarantees its correctness?

Humphrey: I don't guarantee that at all. We must have mechanisms that provide software control. What I guarantee is that there will be nobody "fooling around" moving stuff down into that hardware. I certainly can't guarantee that a yoyo can't program a bad ROM just as easily as he can downline load a bad program. I think that the more critical question is one of software control.

Willard: In terms of the local communications network that you need, assuming that you get as much funding as you really wanted for ISABELLE, what kind of parameters do you need in your local communications network, in terms of node count and total distance and separation between the nodes?

Humphrey: Node count looks to be around 20. The question of distance is one of which philosophy you like. It is possible that I can put all computers in a single room and then communicate over the process data highway to devices. Another possibility is that we put computers out around the ring so that the distance the local computer network has to communicate over becomes larger. The accelerator circumference is 3.8 kilometers, half of that is 1.9 so around 2 kilometers is the maximum distance that I would put on that particular parameter. But once again, I think we will be going out with a request-for-proposal (RFP) for the local computer networking techniques sometime next year. Under the DOE procedures, you go out with a RFP and later you go out for a "request for quote," giving yourself an opportunity to modify your proposals, so that one is a parameter that we probably won't set in the first specification. Once again, this has to do with the philosophy of the vendor, I think.

Sheane: I would like to pursue this device design philosophy a little further, especially on downline loading. Although you won't allow the programs to be downline loaded, have you considered the ability to downline load attributes of the data, at least of the control data?

Humphrey: That, in fact, is what we will allow. For instance, right now we are making a power supply controller that is its own function generator; when you want to change the function as a function of time, there is data that is being shipped out and so that approach will surely be allowed and will be taken advantage of. The important point to be made is that this will have to be done in certain standard ways: all devices will be communicated within a standard way, both hardware and software, both the formats and the way you get the formats down there. These are not just electronics; they are electronics to which we can actually communicate data and we plan to do that. We are spending a lot of time and effort building up microprocessor-based instrumentation.

Sheane: Have you considered the protocol methods and tools that are commercially available now and evaluated them?

Humphrey: That is one of the things that our friends in the Proway specification development are intimately involved in, and it is also one that we will take advantage of. Specifically, one of the topics that has yet to be talked about by the Proway people is the question of the local bus. Those particular discussions are just getting underway. One of things we are using right now is IEEE 488 for communicating with a lot of our microprocessor-based instrumentation. We started that project about three years ago almost when 488 first came out. It seemed like it was a good idea. I think it was a good idea because certainly that standard is particularly well-documented. The solid state industry is supporting it with a large number of very fancy chips. It works a lot better communicating with microprocessor-based instrumentation than CAMAC or other standards that were around at the time. We are using it and making good use of it. We fully expect that IEEE 488 will be used in our control system in some way because there is just a lot of instrumentation that is going to be around, and we are going to have to have it. Questions of whether or not it would be used for some of the local bus loops have to be addressed more systematically. A question of particular importance is its cost effectiveness for such a large usage.

Gellie: I would like to comment on the great value of performance parameter-oriented control, as compared with variable oriented control. This is an important aspect in helping to achieve simplified and improved operation and performance in increasingly complex control systems.

Garcia: In a distributed system a high level language can be the unifying concept which allows the user to create programs in a processor-independent manner. Then a low level module in the operating system would do the translation to create the machine level language. Every time that you bring a new device to become part of the system then you have to have a handler at the operating system to talk to that device. I think that will be one of the tools that would make it a lot easier to have this functional operation concept rather than a parameterized one.

Humphrey: That's exactly the way that the SPS does their system, and it's exactly the way a number of the other accelerators control theirs. In fact, I thought that this was the conversation that we were just having a few moments ago; namely, shipping data and commands down to the devices and those devices know what to do. And it is a similar mechanism to that of the modeling facility, that performs the translation from performance parameters into commands and data that are then communicated to the device itself; it's the data module or device driver with its data base that is the piece of software that communicates between the network and device, to do exactly what you are stating.

Müller: I have three questions: the first one is, since you mentioned CAMAC is not well suited and you are considering IEEE 48 bus, have you in comparison considered others like a microprocessor bus itself, for example, the Intel 8020? Also, what are the reasons for going then to the IEEE bus? The second question is, what are you doing with your control system for things which are safety related? For example, where you have some Boalean equations and where usually you would use in industries these days, a PLC program logic controller unit, are you considering implementing that with a micro-processor? And the third question is a software question: I saw in your paper a short comparison between NODAL and BASIC. What is the intention here at the moment and, if you use an interpretive language in order to make it easy to implement software by a normal operator, are you then considering also a compiled version of that interpretive language?

Humphrey: Our problem with CAMAC is that it is a crate-oriented standard and it doesn't seem to be really the kind of thing that one uses to communicate with relatively autonomous devices. ISABELLE has a lot of this stuff that is not going to go into a crate and so that is our real problem with CAMAC. It is fairly expensive; I think that is fairly well known and has been documented very nicely by Bernard Rausch of the SPS (see references in paper) as an internal note that talks about their cost experience in an environment in which they have a couple of thousand CAMAC modules and seven or eight thousand modules of their own design on another kind of process data highway. Those are the things that went into our thoughts about not using CAMAC. In fact, we are only using IEEE 488 for communications between, not a short distance, but not a long distance either. It only goes up to 10 meters or so, so things that sit in a specific equipment area at ISABELLE, which is an area 17 by about 20 feet, you can cover fairly well with IEEE 488. As to the question of what kind of bus structure to use in an actual device, we are presently using the Multibus. I think the reason for that was that Intel was there first with their development systems and we were into it fairly early and so we took what was available. We are investigating that more carefully but my feeling is right now that it is not a bad decision. In the question of safety oriented things, the answer to your question is "yes," we are thinking about it. We are planning on using some microprocessors, maybe the bit slice type. But, the point is that some of the critical elements of the safety system, those will have to be operated independently of the computer network for lots of reasons. When you are talking about personnel safety and accelerator safety, those are very critical elements. Those have to have a system separate from this kind of system. On the questions of NODAL versus BASIC, for those of you who aren't familar with SPS, one of their features was that they did all their application software in a single language. It was their own version of BASIC that they call NODAL. It takes some things from BASIC, some things from SNOBOL, some things from FOCAL, and mixes it all together in a very nice way for a multi-computer environment. They forced anybody who wanted to program in a high level language to use NODAL, and only NODAL. There is assembly language but all the assembly language coding is done only by Control Systems group members. There is another accelerator system that is being brought on board at the Fermi Labs to control beam lines and they are using BASIC. They are using a compiled version of BASIC also. We like the idea of interpretive languages and we found BASIC quite useful in debugging the systems; especially for people who have not so much computer programming experience. It's a nice way for them to do some quite complicated things. I am unaware of interfaces whereby, for instance, in a FORTRAN program you can call a BASIC subroutine. But surely, if I had a compiled version of BASIC then that problem has been solved, and in fact there are vendors right now that have a compiled version of BASIC available. And so we fully expect to have interpretive language in ISABELLE. It may very well be NODAL instead of BASIC because NODAL has those lovely graphic things which are really quite pretty. The string handling support, and, of course, the multicomputer environment support does make it look very good to us. If we did that, we would have to write a compiler for NODAL.

Gellie: I would like to say one thing about CAMAC. I have long since stopped being a disciple of CAMAC, but I think it's very unfair to say that CAMAC is a crate-oriented standard. There is only one of the CAMAC standards, and there are three or four or five depending on how you count, which is crate-oriented and this is the one which describes the crate and the backplane; the serial CAMAC has nothing to do with the physical packaging and so on. The thing that people confuse is what the vendor is currently offering with what the standard is about. If you are talking about the vendors, the only thing you can get in CAMAC are crate oriented devices.

Humphrey: Thank you for that clarification and you are totally right.

Lalive: I am concerned I do not understand something you said. If there is some DDC which has to be done, it has to be done locally. On the other hand, I think the control from the logic point of view cannot be performed from local information. This is not really true, is it?

Humphrey: What I said is that when I spoke about device design philosphy and said that direct digital control can only be performed locally, I am talking about local devices. In other words, specifically, if a device does not need information in its digital control loop from somewhere else, it gets it all locally. For those kind of devices it's foolish, I think, for the control network to get involved. An accelerator, though, is a big device; it is certainly not local and it clearly needs direct digital control of the accelerator itself. In that case you are taking the information from one place, processing it, and implementing control in another place. That kind of digital control certainly has to be in ISABELLE and is a large fraction of our function, in terms of the control system.

Lalive: That would mean that we just do the control always at the lowest possible level.

Humphrey: That's right, that's exactly the message.

A DISTRIBUTED CONTROL SYSTEM FOR FACILITY ENERGY MANAGEMENT

C. A. Garcia* and V. A. Kaiser**

*Real Estate & Construction Division, IBM Corporation, White Plains, New York, U.S.A.
**Profimatics, Inc., Woodland Hills, California, U.S.A.

Abstract. A new manufacturing, engineering, and laboratory facility for IBM's General Products Division in Tucson, Arizona utilizes a distributed microprocessor-based control system with dual central computers for monitoring and control of energy consumption and space environment. One of the central computers serves as the host for the remote units located in the various buildings on the site, while the other central computer controls central plant equipment and consolidates facility-wide information. The configuration, design, and operational characteristics of the control system are described, as are the on-line scheduling, sequencing, regulation, and optimization functions provided by the system.

Keywords. Distributed control; energy management; facilities engineering; computer control.

INTRODUCTION

The Real Estate and Construction Division (RECD) of IBM Corporation is responsible for providing new manufacturing, laboratory, and office facilities for the company on a world-wide basis. Recognizing the diminishing availability and increasing costs of energy, RECD is placing strong emphasis on the design of energy-efficient facilities and on providing the user of the facilities with the requisite instrumentation and monitoring and control capabilities to permit this energy-efficient operation to be maintained throughout the life of the facility.

One particular example, and the subject of this paper, is the Development Laboratory and Manufacturing Facility now under construction, with partial occupancy, for the General Products Division of IBM in Tucson, Arizona. This Tucson facility will eventually contain some 279,000 square meters (3 million square feet) of floor space, in 19 free-standing structures, with utilities served by a single central plant. This facility was designed for IBM by Albert C. Martin and Associates of Los Angeles, California, and it incorporates a set of design features conducive to a highly energy efficient operation.

PLANT DESIGN

The various buildings at the facility are connected to the central plant by a "spine", shown in Figure 1, which in addition to serving as a covered walkway carries the following services:

- o Chilled water, for space conditioning
- o Heating hot water, for space conditioning
- o Potable cold water
- o Domestic hot water
- o Non-potable water
- o Deionized water
- o Process steam
- o Compressed air
- o Communication cabling

Although electrical power, sewage and industrial waste water are transported separately underground, the "spine" serves as the life blood of the facility.

The entire facility is designed for energy conservation. For example, at the central plant the "stored energy" concept is utilized. Chilled water for space conditioning is generated during off-peak hours with electric-motor or steam-turbine-driven chillers depending on relative operating costs, and the heating hot water is generated, again into storage, as a byproduct of chiller operation.

Another example is the heating of domestic hot water by heat recovery from air compressor operation. Yet another is the selectable use of alternate steam/electric drives for distribution pumps. These and other energy-conscious attributes of the mechanical design are complemented with an equally sophisticated instrumentation, monitoring, and control system which, more to the point, is the real subject of this paper.

DISTRIBUTED CONTROL CONSIDERATIONS

The geographically distributed nature of the facility might lead one to assume that a similarly distributed control system would be appropriate. Such a condition is, however, neither necessary nor sufficient. There are, for example, distributed plants for which a centralized control system is appropriate and, conversely, plants closely-knit geographically for which a (functionally) distributed control system is appropriate.

In the case of this IBM Tucson facility, the significant communication requirements over long distances was a factor, considering the costs of conduit and cabling, but equally important was the relative self-sufficiency required for space conditioning control in the various buildings. In other words, the consequences of a failure in control system components or in communication integrity, can be significantly reduced with a distributed system designed with the proper allocation of intelligence between remote and central portions of the system. In this case, the objective was to keep the building occupants comfortable in the event of a system failure, given that the long-term efficiency of operation is largely determined by mechanical design and cost penalties for short-term sub-optimal operation are minimal.

In the design of any distributed (intelligence) control system, the intelligence dividing line is at the option of the designer. What constitutes a "proper" design of a distributed system is largely determined by the intelligence division; i.e., what degree of autonomy, in terms of function, is retained by the distributed portion of the system relative to that centralized. The designer must assume that parts of the system will, from time to time, fail, and ask himself how he wants that system to perform (and not perform) in that failure mode. This consideration may well be more important than that of initial installed costs; examples can be cited of distributed systems which "saved", relative to initial costs, but cost dearly in operation. Going "distributed" is therefore not the universal panacea but is, with careful design, a viable alternative to be considered in the design of a control system.

SYSTEM DESIGN

The design of the control system for IBM Tucson facility was, as stated earlier, largely based on the objective of maintaining occupancy comfort (or minimizing discomfort) in the event of localized failure. The resulting control system is a combination of "commercial" grade and "industrial" grade instrumentation and controls, analog and digital subsystems, and remote and centralized intelligence. The commercial grade instrumentation and the analog controls are used (1) for isolated backup, in case of failures, and (2) for portions of the control system for which little if any real economic benefit could be projected for a local control system more sophisticated. The industrial grade instrumentation and control and the digital side of the system were, on the other hand, reserved for the portion of the overall system that could be expected to benefit as a result of that more accurate, capable (and costly) alternative.

The IBM Tucson Facility Energy Management Control System is illustrated in block diagram form in Figure 2. At the Central Plant, two IBM Series/1 computers are used; one for the "air side" and a larger one for the "wet side" and the overall site information consolidation. At each building, there is a combination of intelligent "MVCU's" (for Multivariable Control Units) and "LFCM's" (for Lighting and Fan Control Modules) which communicate over separate data highways (two for each building) to (1) the "air side" central computer via a CCM (Communication Control Multiplexer) and (2) an "OPIU" (for Operator Interface Unit) also located in the Central Plant. A second, portable OPIU is provided for local operation and for central plant OPIU backup.

The microprocessor-based MVCU's perform analog and digital input-output and processing operations based on a stored (RAM) applications program which is essentially a configuration of pre-stored (ROM) algorithms. The algorithm library is shown in Figure 3. For a particular application, the program is constructed by assigning input, intermediate, and output variables to these algorithms and creating the algorithm sequences necessary to perform the desired functions.

Each MVCU is fully redundant, with automatic failover controlled by a hardware watchdog timer, and consists of two sets of processor, memory, and input-output control boards as shown in Figure 4. Each "side" communicates on each of the data highways to the central plant, one to the Series/1 computer, via the CCM, and one to the OPIU.

The LFCM's, Figure 5, are designed for digital inputs and outputs only and are not redundant. The MVCU processor board is used in the LFCM, and it communicates on both data highways to the central plant. A dedicated pair of data highways is installed for each building, connecting the MVCU's and LFCM's in the building with the CCM and OPIU in the central plant. Data communication failures should therefore be isolated to individual buildings, and, in case of a communication failure, the portable OPIU can be connected to the data highway in the building for temporary operation and troubleshooting purposes.

The microprocessor-based OPIU's provide facilities for operator configuring, loading, storing, and editing of the applications programs for the MVCU's, and for monitoring and controlling MVCU and LFCM status. These functions can be performed in a conversation mode using a CRT display, keyboard, and two disk storage units. The OPIU's therefore provide a backup mode of operation in case of CCM or central supervisory computer failure, either from the central plant or locally at the building. The OPIU at the central plant is manually switched to one of up to 16 data highways at a time, providing communication to the MVCU's and LFCM's in each building.

The CCM serves as a communication multiplexer between the Series/1 computer and the data highways to the remote buildings, and as a data base storage unit for the remote modules. All MVCU and LFCM status and variable data information is scanned and maintained in memory in the CCM for access by the computer via a RS-232C communication port. Supervisory control and configuration editing commands generated by the Series/1 are routed through the CCM to the appropriate building and local module.

SPACE CONDITIONING CONTROL

A typical administrative or engineering building contains two air handling units, one for the interior space and one for the perimeter space. These units, illustrated in Figure 6, are identical except that the interior air handler has no heating capability. The air handling units supply conditioned air to variable air volume units which operate under thermostat control.

Separate MVCU's control each air handler, through local instrumentation, in the following operating modes:

o Startup
o Warmup
o Flushing
o Normal
o Economizer
o Shutdown

The MVCU will cause the air handler to transfer from one mode to another based on either (1) a stored "mode transfer schedule" or (2) a direct mode transfer command from the central computer. The mode transfer schedule, which can be written to the MVCU from either the OPIU or the Series/1, allows the MVCU to control the air handler continuously according to the present schedule in the absence of communication from the central plant. Separate schedules are maintained for working day, weekend day, and holiday periods, and a calendar keeps track of the current "type" of day for proper schedule selection.

At the times indicated by the appropriate schedule or upon receipt of a direct command, the MVCU performs the actions required to effect the change in air handler mode by modifying its configuration (e.g., opening or closing a three-term discrete controller loop, changing a setpoint of a loop in "automatic" or an output of a loop in "manual") and by direct actions (e.g., starting and stopping fans). The MVCU software organization for mode transfers is illustrated in Figure 7. After performing the mode transfer, the MVCU continues to control the air handler in the selected mode until another mode transfer is made.

Supervisory control from the "air side" Series/1 computer is required to (1) automatically revise the operating schedule and (2) to determine the optimum setpoints to be used for air handler and thermostat control. As an example, the computer evaluates the trade-off between supply air temperature and flow to establish the most energy-efficient setpoint for supply air temperature.

This trade-off arises because of the VAV-box operation by thermostat control: For a given heat load, a change in supply air temperature is compensated for by a counteracting change in supply air volume.

In a normal cooling mode, an increase in supply air temperature (1) reduces chilled water consumption and therefore chilled water generation and distribution costs, but also (2) increases supply air flow requirements and therefore fan operating costs; a decrease in supply air temperature has opposite effects. This characteristic calls for periodic evaluation of the following expression:

$$dC_c = \left(\frac{\partial C_c}{\partial T_{SA}}\right) dT_{SA} + \left(\frac{\partial C_c}{\partial F_{SA}}\right)\left(\frac{\partial F_{SA}}{\partial T_{SA}}\right) dT_{SA}$$

where C_T = Total Operating Costs, \$/Hr

C_c = Chilled Water Costs, \$/°F

C_F = Fan Operating Costs, \$/CFM

T_{SA} = Supply Air Temperature, °F

F_{SA} = Supply Air Flow, CFM

In addition to these air handler control functions, the MVCU collects data for Series/1 evaluation of building and site energy consumption records. The LFCM provides additional digital input-output capability for monitoring and controlling exhaust fans, operating the building lighting, etc.

CENTRAL PLANT CONTROL

At the central plant, a free-standing control panel is provided for manual operation of chillers, pumps, valves, etc. that make up the central plant equipment. These controls can, however, be switched by group from manual to computer control by the "wet side" Series/1. The following groups have been defined:

    Chilled Water Generation
    Chilled Water Distribution
    Heating Hot Water Generation
    Heating Hot Water Distribution
    Domestic Hot Water Generation
    Domestic Hot Water Distribution
    Condenser Water System
    Potable Cold Water Distribution
    Non-Potable Water Distribution
    Deionized Water Distribution

In general, the Series/1 computer monitors central plant equipment operations, provides logs, alarms, and performance reports, and controls the sequence of operations of the equipment. Computer control is at the discretion of the plant operator, who may select computer or manual control via switch operations at the control panel or CRT console entries. This selection may be made independently for any of the ten functional plant areas listed above. Computer control of a portion of the plant is compatible with manual control of the remainder; operator requests, rather than direct control actions, are provided for the operation of equipment under manual control.

Use of the Series/1 Event Driven Executive (EDX) operating system facilitates the programming of the central plant monitoring and control functions, and allows the use of a convenient format for specifying the logical sequences required and documenting the applications software design. By way of example, the control for the chilled water storage tanks is described in the following paragraphs.

The chilled water storage system consists of nine 300,000 gallon insulated tanks, connected in series as shown in Figure 8. During regeneration, chilled water is added to storage by routing the chiller product to the highest numbered tank not yet filled with chilled water; the remaining tanks (except for an equivalent of one tank, which is always empty) contain the warmer chilled water returned from the buildings. When sufficient chilled water has been generated into storage to meet the following day's requirements, the chillers can be stopped and the chilled water supply, for building space conditioning, can be drawn from storage.

In either mode of operation (generation and withdrawal) there is only one active tank (being filled or emptied) at any one time. The control logic must therefore monitor that "active" tank and, when it becomes full or empty, operate the valving to make the next adjacent tank the "active" tank. The valve operating sequence depends on which "mode" the system is operating in; the two mutually exclusive modes are:

    Mode 1--When total chiller flow exceeds chilled water demand, so a net regeneration or replenishment of chilled water inventory is taking place.

    Mode 2--When total chiller flow is less than chilled water demand, so a net draw or depletion of chilled water inventory is taking place.

The valve switching sequence for Mode 1 operations is illustrated in Figure 9. When the Nth tank being withdrawn from becomes nearly empty, or the adjacent (N+1)th tank becomes full, the valves are switched so that withdrawal is from the (N-1)th tank. The valve switching operation is performed in three steps as illustrated:

Step 1: The (N-1)th tank valve is opened.

Step 2: The (N-1)/N line valve (i.e., the line valve between tanks (N-1) and N) is closed.

Step 3: The (N-1)th tank valve is closed and the N/(N+1) line valve is opened.

If the chilled water flow to storage (i.e., into the Nth tank after switching) is sufficient to minimize the possibility of a mode reversal (e.g., an increase in demand which could cause withdrawal from the Nth tank), a fourth step in the valve sequence is performed:

Step 4: Close Nth tank valve.

This fourth step causes the regenerated chilled water to overflow the higher-numbered tanks into the Nth tank rather than flow directly into the bottom of the Nth tank. This step is reversed, however, if chilled water flow to storage is reduced to allow the chilled water distribution pumps to withdraw from the Nth tank if required.

The Mode 1 valve sequencing logic is summarized in tabular form in Figure 10, and an example of this logic presentation for the Series/1 EDX program entry is illustrated in Figure 11.

The valve switching sequence for Mode 2 operations is illustrated in Figure 12. In this mode, when the Nth tank being withdrawn from becomes nearly empty, or the adjacent (N-1)th tank becomes full, the valves are switched so that withdrawal is from the (N+1)th tank. The valve switching operation is performed in three steps as illustrated:

Step 1: The (N+1)th tank valve is opened.

Step 2: The N/(N+1) line valve is closed.

Step 3: The (N-1)th tank valve is closed and the (N-1)/N valve is opened.

Because mixing of the return water is not required, a fourth step is not required as it was for Mode 1 operations. This Mode 2 valve sequencing logic is illustrated in tabular and in logic diagram form in Figures 13 and 14, respectively.

This valve sequencing for chilled water storage tank management is one of the simplest operations performed by the computer. Much more complicated logic sequences are required, for example, for chiller control when alternate electric vs. steam chiller drives and pumps must be selected, and condensing water pumping and valving and cooling water operations must be coordinated with chiller startup, operation, and shutdown.

SUMMARY

The Facility Energy Management System at the IBM General Products Division plant in Tucson, Arizona, provides a specific example of a specially-tailored distributed control system designed to meet particular objectives. The functions to be performed by the system and the effects of a potential component or subsystem failure were major considerations in arriving at the final design and allocation of functions among the distributed elements.

From an energy management point of view, much of the "optimization" is inherently incorporated in the mechanical design, leaving well-defined functions for the control system to perform to further energy conservation on a day-to-day basis. It is to be expected, however, that economic and operating conditions for the facility will change over time, necessitating changes in both mechanical design and control system design to adapt. It is also expected, however, that the structure, configuration, and system module capabilities of the original control system will be able to accommodate these changes as they occur in the future.

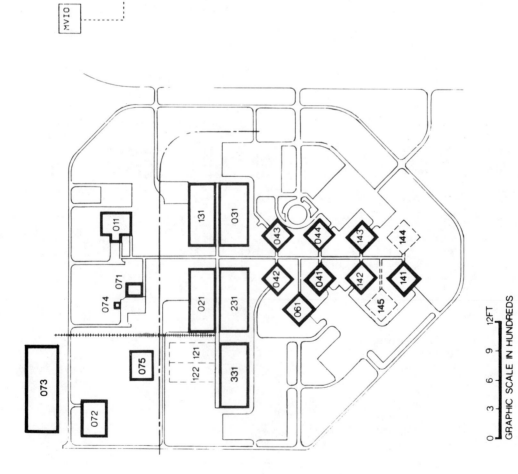

Fig. 1. Site plan.

Fig. 2. Control system configuration

System for Facility Energy Management

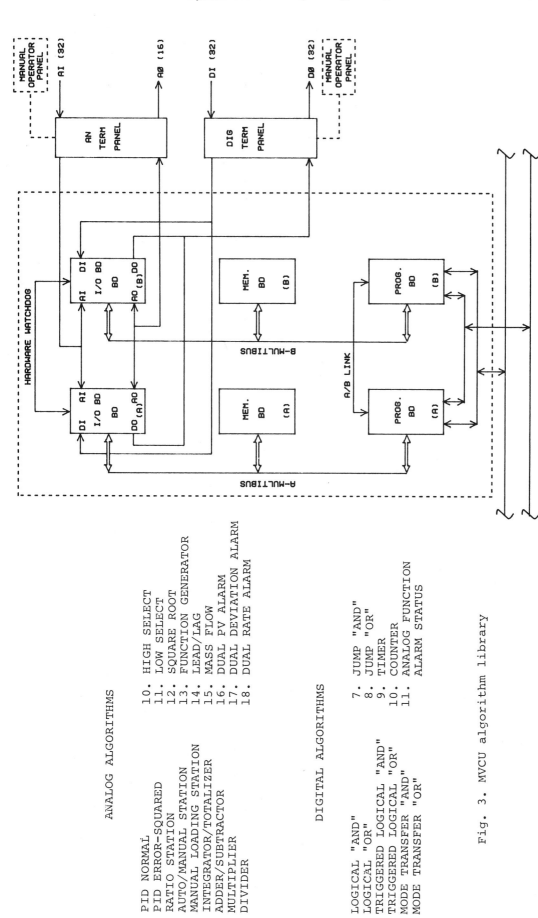

Fig. 4. MVCU board layout

ANALOG ALGORITHMS

1. PID NORMAL
2. PID ERROR-SQUARED
3. RATIO STATION
4. AUTO/MANUAL STATION
5. MANUAL LOADING STATION
6. INTEGRATOR/TOTALIZER
7. ADDER/SUBTRACTOR
8. MULTIPLIER
9. DIVIDER
10. HIGH SELECT
11. LOW SELECT
12. SQUARE ROOT
13. FUNCTION GENERATOR
14. LEAD/LAG
15. MASS FLOW
16. DUAL PV ALARM
17. DUAL DEVIATION ALARM
18. DUAL RATE ALARM

DIGITAL ALGORITHMS

1. LOGICAL "AND"
2. LOGICAL "OR"
3. TRIGGERED LOGICAL "AND"
4. TRIGGERED LOGICAL "OR"
5. MODE TRANSFER "AND"
6. MODE TRANSFER "OR"
7. JUMP "AND"
8. JUMP "OR"
9. TIMER
10. COUNTER
11. ANALOG FUNCTION ALARM STATUS

Fig. 3. MVCU algorithm library

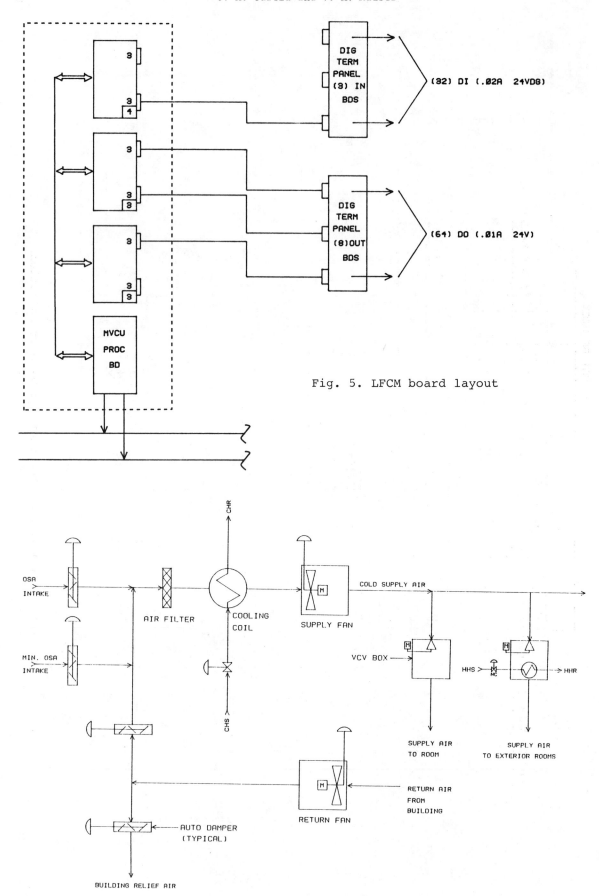

Fig. 5. LFCM board layout

Fig. 6. Typical air handling system

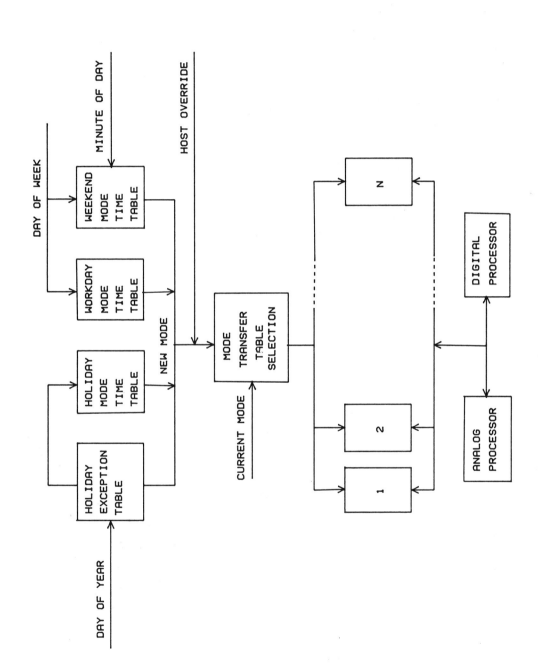

Fig. 7. Air handler mode transfer software organization

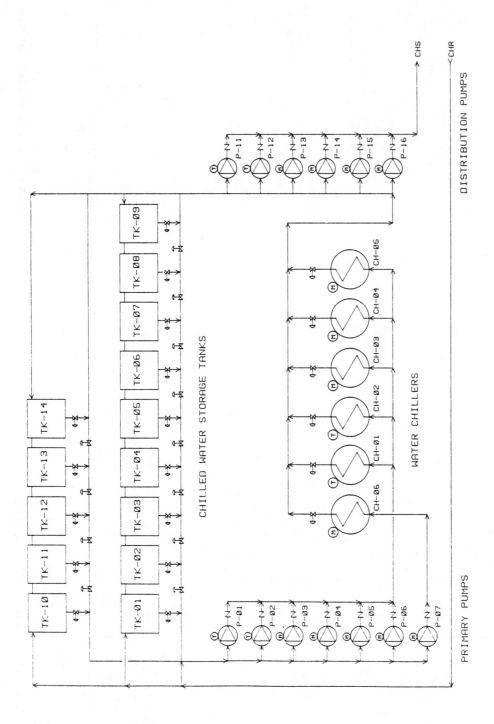

Fig. 8. Chilled water storage system

System for Facility Energy Management

IF	AND	AND	THEN	
8 OPEN	8/9 CLOSED	L8 < LL OR L9 > HL	OPEN 7 CLOSE 7/8 OPEN 8/9 CLOSE 9	(DO13) (DO32) (DO33) (DO18)
7 OPEN	7/8 CLOSED	L7 < LL OR L8 > HL	OPEN 6 CLOSE 6/7 OPEN 7/8 CLOSE 8	(DO11) (DO30) (DO31) (DO16)
6 OPEN	6/7 CLOSED	L6 < LL OR L7 > HL	OPEN 5 CLOSE 5/6 OPEN 6/7 CLOSE 7	(DO 9) (DO28) (DO29)
5 OPEN	5/6 CLOSED	L5 < LL OR L6 > HL	OPEN 4 CLOSE 4/5 OPEN 5/6 CLOSE 6	(DO 7) (DO26) (DO27) (DO12)
4 OPEN	4/5 CLOSED	L4 < LL OR L5 > HL	OPEN 3 CLOSE 3/4 OPEN 4/5 CLOSE 5	(DO 5) (DO24) (DO25) (DO10)
3 OPEN	3/4 CLOSED	L3 < LL OR L4 > HL	OPEN 2 CLOSE 2/3 OPEN 3/4 CLOSE 4	(DO 3) (DO22) (DO23) (DO 8)
2 OPEN	2/3 CLOSED	L2 < LL OR L3 > HL	[STOP CHILLERS]	

Fig. 10. Mode 1 valve switching summary

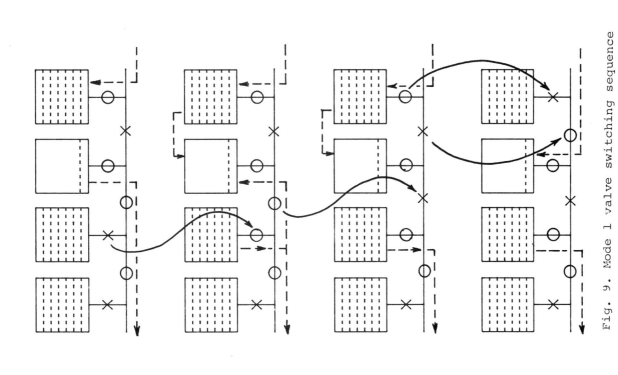

Fig. 9. Mode 1 valve switching sequence

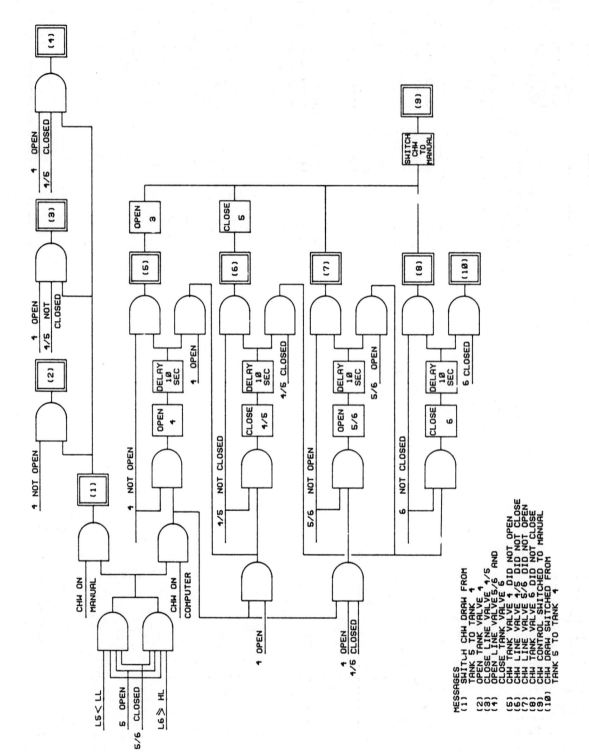

Fig. 11. Typical Mode 1 logic diagram

System for Facility Energy Management

IF	AND	AND	THEN	
2 OPEN	1/2 CLOSED	L2<LL OR L1≥HL	OPEN 3 CLOSE 2/3 OPEN 1/2 CLOSE 1	(D0 5) (D022) (D019) (D0 2)
3 OPEN	2/3 CLOSED	L3<LL OR L2≥HL	OPEN 4 CLOSE 3/4 OPEN 2/3 CLOSE 2	(D0 7) (D024) (D021) (D0 4)
4 OPEN	3/4 CLOSED	L4<LL OR L3≥HL	OPEN 5 CLOSE 4/5 OPEN 3/4 CLOSE 3	(D0 9) (D026) (D023) (D0 6)
5 OPEN	4/5 CLOSED	L5<LL OR L4≥HL	OPEN 6 CLOSE 5/6 OPEN 4/5 CLOSE 4	(D011) (D028) (D025) (D0 8)
6 OPEN	5/6 CLOSED	L6<LL OR L5≥HL	OPEN 7 CLOSE 6/7 OPEN 5/6 CLOSE 5	(D013) (D030) (D027) (D010)
7 OPEN	6/7 CLOSED	L7<LL OR L6≥HL	OPEN 8 CLOSE 7/8 OPEN 6/7 CLOSE 6	(D015) (D032) (D029) (D012)
8 OPEN	7/8 CLOSED	L8<LL OR L7≥HL	OPEN 9 CLOSE 8/9 OPEN 7/8 CLOSE 7	(D017) (D034) (D031) (D014)
9 OPEN	8/9 CLOSED	L9<LL OR L8≥HL	OPEN 8 OPEN 8/9 [ALARM]	

Fig. 13. Mode 2 valve switching summary

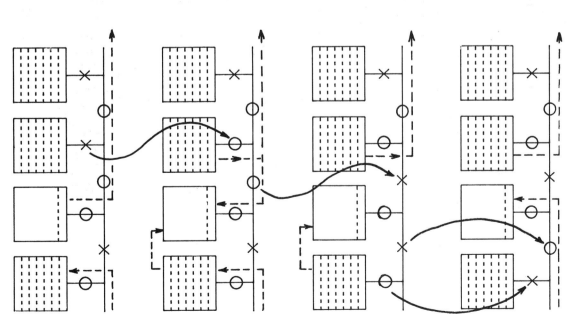

Fig. 12. Mode 2 valve switching sequence

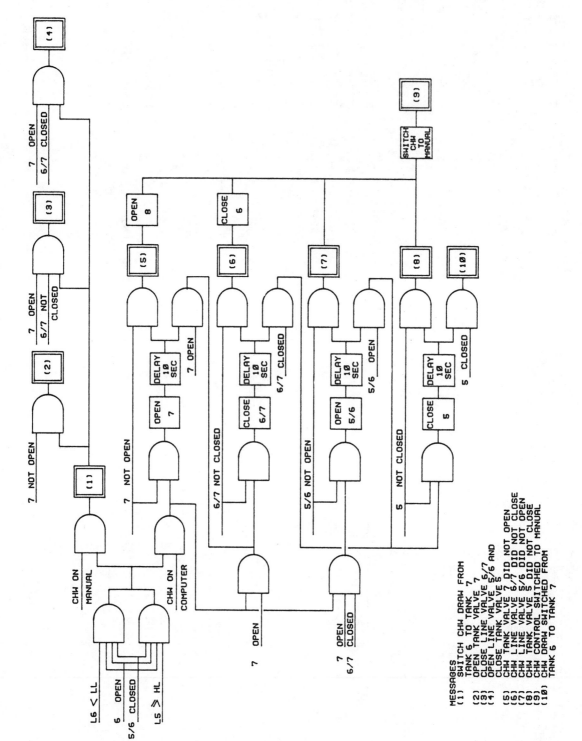

Fig. 14. Typical Mode 2 logic diagram

DISCUSSION

Gellie: Do I understand that this system, at the moment, controls just the environment in the building, or are plant demands tied in with the manufacturing demands for peak loading considerations? Are there plans for overall energy management in this sense?

Kaiser: The coordination and control of site power demand will all be done at the central plant.

Lalive: To me it seems obvious that there should be a link between the operator console and the central Series/1. Why is there no link?

Kaiser: Well, several reasons: The Series/1 and the OPIU were supplied by different sources and were expected to arrive on site at different times. The purpose of the OPIU is for initial configuring and loading of the MVCU's, and for backup communication in case the Series/1 is not available. There is also a CRT console for communication to the operator through the Series/1, so the OPIU was really intended as a backup and system startup device.

Lalive: If it is for backup and if one of the bus systems or communication links fail, then you will not be able to support the system from your central machine. To use the other bus for backup purposes, such a redundant link would be required. If the bus between the central plant and the buildings fails, then you cannot use the central computer.

Kaiser: What you say is true and a redundant link could be added; there is no difference between the signal going to the CCM and the signal going to the OPIU from the buildings. It's a matter of rewiring and switching. But, what we have done to get around that problem is, as I have mentioned, to have a portable OPIU unit that you can take up to the building. So, if you lose the link between the central plant and the remote units, you can communicate with the MVCU's directly at the building.

Singer: Do you "zone" within each of the smaller laboratory buildings and handle those with the ability to control small areas of the buildings on an individual basis?

Kaiser: I believe there are eight zones on each system. The exterior air handler has eight zones and the interior air handler has eight zones.

Humphrey: Back to what Dr. Lalive was talking about, you have two operator interfaces; how do you keep only one up? That is, how do you prevent the possible conflict from two different operation consoles? For example, what if two operators both try to control the same building?

Kaiser: There is a lock-out mode in the software at the MVCU. From either the OPIU or the Series/1 location, you can lock out communication on the other bus while you are reconfiguring, for example. But typically, we don't expect that to happen because we placed this OPIU console behind the main control panel in the central plant. It's just for backup.

Humphrey: So, you are saying it's in the same location?

Kaiser: The console for the Series/1 computer is located in front of the control panel, where the operator normally sits. The OPIU is less accessible behind the control panel, and will be used normally just for backup.

Hyde: Can you talk about how much some of those pieces cost, how much it costs to heat a building like this, what is your marginal return on a building which is without centralized control, and what the centralized control returns when you get it on line?

Garcia: We are not authorized to talk specific cost figures; however, the guidelines under which we operate for a system like this is that any cost, over and above what a conventional control system would be, has to be justified on a return-on-investment basis. Therefore, that constraint is there; we cannot go freely with the design of the system like this.

I would like to add one comment on the perceived benefit of a system like this from the user perspective, and this is something which we have found to be very important. When we previously implemented these kinds of systems with centralized computer control, as opposed to a distributed system, the programming considerations always were of an overwhelming nature when compared to the control considerations. For a long time during the design and implementation of a system, we could never get out of the software element of the project to start talking about the application. With a system like this, which is fairly simple in its architecture, we were able to concentrate on the control problems at a very early stage in the project and, what used to be significant software problems, such as control module independence, linkage of control modules, and operating system related problems, all disappeared with the fact that we physically separated the control functions and placed them in an independent location. I think this was, perhaps, one of the greatest benefits that we saw from a distributed system for this particular kind of application; it gave the user the control of the problem again and took it away from the programmer. I think that was very important.

Harrison: Although we can't discuss the specifics of Tucson, I can tell you that System/7 and Series/1 computers are in operation at 62 IBM locations to control and minimize the use of energy in the facilities. Based on the company's experience, we believe that computer control can provide additional energy savings of 10-15 percent at many of the locations where installed.

At the end of 1978, almost 900 IBM customers were using computers to control and reduce energy use. Many have reported savings of up to 25 percent in electricity. IBM's total projected savings of its customers is more than 2.5 billion kilowatthours annually.

Gellie: We've heard two widely different applications described this morning. As well as asking questions about this paper specifically, I would also like to encourage discussion on the two papers and the similarity and the differences between the two different problems. They are very different. They are both real time, in a sense, but the definition of real time is sort of night and day between the two of them. Also, the consequences of failure is rather different between them. If the room suddenly goes up to 75 degrees, it's not quite the same as losing your beam in ISABELLE. I would like to encourage comparison and discussion of topics of mutual interest to the two papers.

One, in particular, that I would like to suggest is the question of what are the criteria for deciding on the allocation of functions in the topology that you choose for a distributed system. I have always had a gut feeling that one of the criteria for a good topology and distribution of function is that you have minimized the communication that actually is required to keep the system glued together.

Kaiser: I would like to comment on that. I was reading the abstracts of the workshop papers relative to the theory of distributed systems and I am curious to hear those papers, because it seems to me that topology is so application dependent; it would almost be impossible to develop a unifying theory. As you mentioned, one of the key things is the allocation of function and we've seen some pretty terrible mistakes made in these kinds of things. I can give you a couple of examples; one was in a system similar to this where the controlling computers were separated by some distance, but they were controlling, essentially, either ends of related equipment like a conveyor that feed onto another conveyor. Startups and shutdowns had to be performed in the right sequence. With one end controlled by another, the only way that you could interlock the two conveyors, for example, was to communicate back to the central computer and back out to the other computer. Now, that's a real fair weather companion because everything has to be working for anything to work! That's an obvious mistake, of course but it points out the application dependency of the system configuration. Is that what you were referring to?

Gellie: Well, that is certainly one of them. I'm not sure whether they are really application dependent or whether there is some basic common sense rules that underline all of them. It would be very useful if somebody could list and catalog some of these basic "do's and don'ts."

Willard: Let me conjecture that in many instances in the real world, the optimal allocation of functions to processes in a multiprocessor system, as distributed control systems are, really is frequently secondary to the issue associated with ease of startup and ease of phaseover from a previous control system to the new control system. I think that is one of the guiding issues in the IBM Tucson facility.

Gellie: Yes, I think that's true. But that's what engineering is all about, isn't it? It's making the tradeoff between the ideal and the practical and you know one should always go for the ideal until you have a good justification for going away from it. To just say, " we did this," without realizing that you didn't

Discussion

do what was best, and the reason you did it was because of the engineering reasons for the tradeoff, that's the important part!

Lambert: I have two questions, the first one is very quick. Is the OPIU another Series/1?

Kaiser: No, sir. Except for the two Series/1 computers, all equipment shown was supplied by Beckman Instruments. The OPIU is a microprocessor-based terminal, terminal, specifically designed for communication with the MVCU's. It has the conversational configure, edit, and read modes for operating with this remote equipment.

Lambert: The second question is directed to Carlos Garcias. I was interested in his comment. I think I am summarizing it correctly, that you avoided a lot of the software difficulties because of the distributing processing concept that you use. In other words, you interjected simplicity into the development process. That's interesting; could you venture a guess as to what the percentage of software content was in this implementation as opposed to an implementation you may have done in a different manner? What I am saying is that in a lot of cases in the past, 60% of the effort, or better, has been software effort as opposed to hardware content. I think you alluded to the fact that maybe you got a much better ratio this time. Is that what you meant by that statement?

Garcia: It is difficult to quantify, in the sense that there is software down at each one of the modules, the microprocessors. But what happens is that, when you look at each one of those modules independently, that piece of software is very simple and straightforward. Essentially, it consists of two elements: One is the application program itself and, second, a communication module that talks back to the central system. So when you start doing that element of the software, you are already dealing with the application. Now, when you are operating on a central computer control basis, that is the last thing you get to since you have already gone through all the complexities of dealing with a rather sophisticated operating system and a lot of software considerations which are independent of the application. So, that has been the beauty of a system like this: practically from day one, everybody involved, including the software people, were talking application. That is very important.

Lambert: I understood what you said. I was hoping, against hope I guess, that you had some quantitative data that said this is the way to go if you want to improve your software productivity and, more importantly, improve your software quality. Doesn't anybody ever keep any data like that? Obviously, you have implemented systems like this before in the past. Was the quality of the software produced better? In other words, did you get to a stable system quicker than you have with a nondistributed system?

Garcia: The answer is "yes." I cannot give you quantitative numbers, but during our limited experience in the start-up of his system, we had some minor software glitches that we were able to trace in two to three hours. Because of the ability of isolating components so very quickly, we knew that the problem was at a particular box; the total software in that box was very small, so you could go through it in no time and find out what the problem was. If you have that kind of a problem in a centralized system, it could take weeks before you could find out what little thing was giving you a problem.

Kaiser: I think, in general, that there is a relationship between complexity and cost which is probably close to exponential. By making systems modular and simple, the modular devices, and the software for those devices, can almost be rubber stamped; whereas, on the other hand, if a single system was designed to do everything, then your subroutines and indexing and so forth can get quite complicated.

Lambert: I understand what you are saying, but all I get is "gut feels" and, when you are trying to determine return-on-investment for your next project, it would be nice if you had quantitative data that the distributed processing approach with microcomputers gives much better quality or that much better productivity. I certainly think that should be very important to Rusty Humphrey; on his project, quality is probably even more important.

Humphrey: My numbers come from the SPS experience and they are applicable, but I'm afraid they are not going to be exactly what you want, because we only build accelerators once ever five to seven years or something like that. So, everyone faces new technology each time. My favorite numbers out of the SPS control system are that they have two hundred thousand lines of application code (NODAL code) and 60 different people that write NODAL code. These are not people in the control systems group. By any reasonable measure, that is probably something of the order of 100 manyears of effort that the computer systems group was not responsible for. There is a roughly similar amount, something like 55-60 man-years of control systems software effort, in the building of that particular facility. Now, the other important parameter is that the SPS control system worked when they turned the accelerator on and that was the first time that has ever happened!

Steusloff: I am wondering why you don't have a link between the operator interface and the central plant control. If I understood it right, the operator interface is some kind of backup if the central system fails. It should also be used for operating backup to the central plant if there is no communication from the central plant.

Kaiser: That consideration was taken care of in the central plant in a different way. The central plant has a full control panel, with instruments, indicators, controllers, and so forth; the objective was not to do anything with computer equipment that would keep the operator from manually operating that plant, such as running the chillers, the pumps, and so forth. The computer system is just there to assist him. He has all the capability he needs for control without the computers and he is there 24 hours a day. The fan rooms at the buildings areas are unmanned, so we had to provide the backup facilities there. At the central plant, the control panel is the "backup."

Lalive: I have a question about the dual redundancy which is in the local systems. Normally, redundancy increases reliability, but it also increases the number of maintenance cases. You said that you would reduce the number of times people have to go to these remote stations. Does that mean that if one system fails you continue to run until the other ones fail? That would be my conclusion.

Kaiser: It buys you some time; that is what it does.

Lalive: Is the central system notified if one of the two systems fails, and what happens if the standby unit fails? Do you notice that?

Kaiser: Yes, the central system monitors both the status of each bus and the status of each side of each MVCU. As a matter of fact, all of that information is available at the OPIU and at the Series/1, as well. We have a display which shows bus status, MVCU status and so on. To be specific about failure and repair rates would be difficult because we don't have enough operating experience in Tucson, yet, with this application. In general, however, I would expect that the redundant MVCU side really buys time to get service out to repair the side that malfunctioned.

Merluzzi: I am still a little confused about the software saving that Mr. Garcia spoke about, in the sense that he was talking about operating systems programming on one hand and the configuring of the microprocessor units on the other. Aren't you really comparing apples and oranges, because many of these preprogrammed algorithms can also be readily available, say, on a central processing unit and linkable in the same way that you would on a microprocessor controller?

Garcia: That is true, but what happens, and that phenomenon we have seen, is that with the rapid changes in generations of computer hardware, we are always playing a catch-up game between the systems software and the application. Where a new generation of computers is available and you have to start doing a lot of work on systems programming before you can put the first application on line, by the time that systems software is really tested and in good operating condition, the new generation is coming along. So, with this approach you are deemphasizing the systems software aspects and giving more emphasis to the applications software. I think that is something that is very useful in this particular application where, as we saw before, some of the timing considerations, and other system considerations, are not as demanding as they would be in the case of, say, ISABELLE. So because of that, we have been able to move into the application area quicker. But you are correct; if we were going to freeze these types of applications with the same kind of software, in the long run it probably would not make that much of a difference.

PROGRAMMING DISTRIBUTED COMPUTER SYSTEMS WITH HIGHER LEVEL LANGUAGES

H. U. Steusloff

Institute for Information and Data Processing, Fraunhofer-Gesellschaft, Sebastian-Kneipp-Str. 12-14, D-7500 Karlsruhe 1, Federal Republic of Germany

Abstract. During the past few years the increasing demands for system automation, together with the progress in VLSI-semiconductor technology, have introduced distributed computing systems, yielding the potential of higher performance and higher availability. Those hardware properties have to be accompanied by suitable programming aids to make them explicitely usable for the application programmer. In addition, software costs have to be maintained at the same level as the software costs for centralized systems.

Our approach to suitable programming aids starts from two theoretical assessments using systems theory and decision theory. The requirements and the design of a higher level application programming language for distributed systems are derived from these assessments. The realization of this language design employs a value analysis of existing languages to determine the best suited candidate for the basic requirements. The language chosen is PEARL (Process and Experiment Automation Realtime Language) which has to be extended for the requirements of distributed computer systems. These extensions - not altering basic PEARL itself but adding new language elements - are described in some detail together with supporting system programs. The PEARL-extensions provide means for the description of hard- and software structures and/or configurations, as well as the explicit formulation of exception handling, i.e. reactions on non-regular system-states, where the "system-state" is given by the actual values of a set of system-variables. Finally this language is classified into a hierarchy of language levels ranging from very high languages to system development languages.

Keywords. Process control; computer software; decision theory; distributed multicomputer systems; microprocessors; programming languages; exception handling; redundant systems.

INTRODUCTION

Consideration of the programming of distributed computer systems should start from a discussion of distributed computer systems structures and architectures. Changing the structure of computer systems, we feel, is the only long term qualitative answer to the quantitative challenge of increasing requirements and complexity in process automation. Changing the structure of computer systems means overcoming a philosophy expressed by Grosch's "Law" (Knight, 1966) which calculates the efficiency of a computer system by the expression

$$costs \sim a \cdot \sqrt{capacity}$$

and thus favouring the large central computer. This was acceptable as long as the capacity of a computer system was only determined by its hardware, where it was combined with high costs. Today, in the face of rapidly falling hardware costs by the VLSI-technology, as well as prevailing software expenditures, Grosch's "Law" can no longer predominantly justify the use of large centralized main frame computers.

Further problems arise with the increasing demands of high availability and, on the other hand, security of process computer operation. Here new solutions have to be found for the domains of maintenance and transparency of system characteristics and behaviour, even in the case of disturbances. It will be shown in this paper that distributed systems offer outstanding properties for efficient solutions to these problems.

To investigate the optimal way of changing the structure of computer systems in general, and especially process computer systems, it is necessary to enquire as to the structure of the technical or commercial process to be automatically controlled. Knowing this structure, it is possible to derive the optimal structure of the related process automation system. Furthermore, if we consider programming languages as a means for the description of process automation systems, we can determine the language features of a suitable application programming language, bearing in mind the requirements for efficient utilization of this language. This paper will show the way to, and the realization of, such a process automation language.

OPTIMAL STRUCTURE AND CAPABILITIES OF AN AUTOMATIC CONTROL SYSTEM

The question of the optimal structure of an automatic control system corresponds to the question of the structure of the technical process to be automatically controlled. From systems theory (Kalman, Falb and Arbib, 1969), methods for the description of technical processes are known, by means of which the following duality principle may be formulated (Syrbe, 1978):

> An automatically controlled technical system (technical process) consists of distributed, parallel operating, coupled, partial processes.
> The dual controlling system consists of at least the same number of corresponding, parallel, equally coupled processes (in the sense of information theory). Hardware and/or software are thereby determined.

From this duality principle, the optimal structure of the optimal control system for each controlled technical process is basically determined. Figure 1 shows the three basic types of components of automatic control systems in a general network, which may be called polyhedron. These types of components,

- the field-units (sensors and actuators),
- the processing components (computers, controllers), and
- the coupling- or communication components,

may, in general, also be combined leading to the question of different alternatives for the realization of such structures.

Starting from the polyhedron as the basic optimal structure, several distributed process automation system structures can be derived (Steusloff, 1979). By the way, even the centralized process computer system contains an intrinsic polyhedric structure by mapping the polyhedron to a set of sequentially running tasks which communicate via a common data base.

Since this paper is dedicated to the programming of distributed systems, we will not discuss the several possible structures. To show the properties of distributed process automation systems such as

- decrease of cabling expenses,
- decrease of the complexity of tasks in the distributed processors,
- increase of comprehensibility and improved modularity of automatic control systems,
- increase of availability with simultaneous increase of safety of critical control functions,
- economic realizability of these properties,

a system will be presented, which has been realized and is now in operation for the automation of a soaking pit plant in a big German steelworks (Heger, Steusloff and Syrbe, 1979). Prior to explaining the system, a special approach to increased system availability shall be introduced.

Function-Sharing Redundancy

In distributed automatic control systems, due to the duality principle, there exist several processing components and a communication system, often with several alternative ways for information exchange. For each of these system components there are also redundant components, if according to the design of the system for normal operation, the capacity of each component is over-dimensioned in the necessary way. This type of redundancy is called functional redundancy or more precisely function-sharing redundancy.

While dimensioning stand-by redundancy is achieved by multiplying the normally required number of components, function-sharing redundancy may be dimensioned according to the problem and thus may be optimized economically. For this purpose, the functions of an automatic control system are subdivided into classes of different importance, such as

I   Indispensable functions (e.g. safety functions),

II   Parametrizable functions (e.g. decrease of sampling-rate of control-loops),

III   Temporarily dispensable functions (e.g. printing of stored balancing data, programming).

The individual system components have to be dimensioned in such a way that after troubles, functions III are dispensed with first.

Free system capacity is now utilized by shifting of programs or switching to alternative data ways. After further faults, functions II are switched to parameters requiring less system capacity. Finally, enough components (operating means) remain for functions I. In this way a range of system performance is achieved, called "graceful degradation", which permits, among other things, scheduling of repair of faulty components. The programming language for such a system must provide means for the description of its capacity and the utilization of the system-inherent function-sharing redundancy.

As an example of a distributed process control computer system with decentralized communication, Fig. 2 shows the general outline of a system now in operation at a big German steelworks. As a basic unit, the so called Micro Computer Stations (MCS) carry out the "local" automation of the partial processes. Each MCS is equipped with a process control microprocessor (PµP), a set of process I/O-devices, adapted to the requirements of the respective partial process and a communication microprocessor (LµP). As a central unit of each MCS, Fig. 2 shows an internal bus-switch with the ability of connecting or separating the three internal partial busses to the I/O-, the PµP- and the LµP-unit. This bus-switch unit (BSU) also contains the MCS error detection unit and switches the partial busses according to the actual error conditions. In general there are three possibilities:

- The PµP is out of service.
  The BSU connects the I/O-bus to the LµP-bus to make its I/O-values available to other MCS's via the communication busline.

- The LµP ist out of service.
  The BSU disconnects the LµP-bus and the MCS now is working in an isolated status.

- The I/O ist out of service.
  The BSU cuts off the I/O-bus. If the process field devices have been connected to the I/O of different MCS's (dotted lines), the PµP-unit is able to access them via the communication busline.

In the latter system state, there is a high data transmission load on the communication busline, as it will be during the reload phase after the repair of a broken-down MCS or during software reconfiguration time, following the detection of an error in one of the MCS's. The necessary high transmission rate on the communication busline has been achieved by an optical fibre line with a 1.0 megabytes/sec transmission rate and a ring-shaped structure. Function-sharing redundancy is established by this powerful communication link together with error detection equipment in each MCS, alternative data ways and over-dimensioned storage capacity, as well as some spare processor-time of the micro-CPU's.

There are two special computer stations in the system. The first one (lower right hand side in Fig. 2) is used for process-operating and is equipped with two colour-screen-input/output-systems, employing light pens and virtual keyboards for the command input. On the lower left hand side, Fig. 2 shows the programming and documentation system. Both systems are connected by communication links and cross-over-switches for the periphal units. The programming system serves as function-sharing redundancy for the process-operating system.

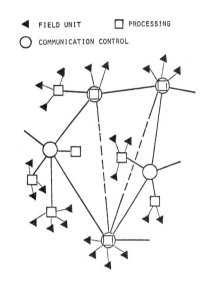

Fig. 1. Polyhedron Structure

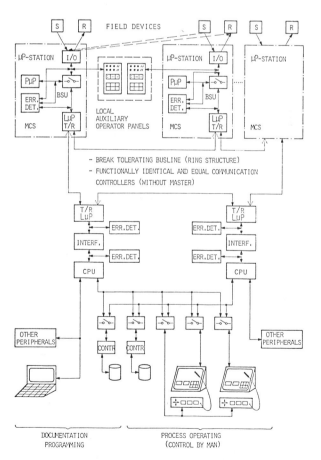

Fig. 2. Distributed Process Control Computer System

APPLICATION LANGUAGE REQUIREMENTS FOR DISTRIBUTED COMPUTING SYSTEMS

In the past, the formulation of processing <u>functions</u> in a programming language, mostly assembler, was predominant in application programming of computer systems. This was sufficient, as long as the hardware structure of the computer system was fixed, except for a variation of quantity parameters (storage capacity, number of I/O units etc.). Changes in hardware structure after system faults either meant total failure of the system or switching to identical redundancy, or had to be treated by the program as variants of the processing functions (e.g. processing of error messages of the I/O-channels). For complex computer systems like automatic control systems (e.g. in a star structure and with high availability) the error processing (exception handling) parts of the programs are often larger than the actual processing functions. A further part of error processing is frequently integrated into the operating system of the process computer and cannot be influenced by the user. Reactions to such errors often have to be formulated in another language, the control language of the operating system.

Distributed computing systems with function-sharing redundancy are subject to structural variations due to extensions and to their actual operating state. Therefore the programs have also to describe - besides the processing functions - the structure of the system and its communication. The advantages of function-sharing redundancy can only be exploited if <u>all</u> significant states and structures of the system, as well as the corresponding exception handling, are explicitly described in user programs. It should be formulated, easily recognizable, and in <u>one</u> programming language, under which operating conditions and states the distributed computing system exhibits a certain capacity. In the same way, it must be made clear, for which operating states there is no redundant system capacity available and how large the degradation in performance may become in those states. Within a uniform language design, the software aids for structural description should be separated from the means of describing processing functions, in order to make programming of distributed automatic control systems not substantially more complicated than that of centralized systems.

Starting from these general statements we will now derive - by a "top-down" method - the requirements and the design of an application programming language suited for real-time distributed computing systems.

Description of Distributed Computing Systems

Considering an application programming language as means for the description of an abstract computing system, defined by the semantics of the programming language, first we have to ask for an appropriated description method for these systems. Several methods have been proposed, one of them (Leinemann, Schumann, 1973) giving a general subdivision of the features of a system into

- properties,
- functions,
- structure.

All the objects of a technical system can be described by such a tripel, where "object" comprises all parts of a system, programs and programming aids included. In some cases there are special parts of a system's description given implicitely or by convention and thus being a "standard prelude" which can be added to the actual systems description. These "preludes" have been of great importance in programming languages and, within single computer systems, usually comprise the description of properties and structure. Distributed and heterogenous computing systems today enforce the explicit description of many more features, and the need of suited programming language semantics and elements arises.

To investigate the necessary elements of a real-time application programming language in the "top-down" way we formulate a principle of duality between real-time automation systems and programming languages (Steusloff, 1977):

> For each partial process there exists at most one section of the automatic control system with the same <u>features</u> as the corresponding partial process. These features of the automatic control system have to be described by a programming language. The software production system for this language will map these features to a computer system.

Figure 3 shows this correspondence between a continuous technical system and a discrete computer-aided automation system. From the discrete version of the system equations it is possible to extract the necessary features to be formulated by use of a programming language. Then the corresponding language elements can be derived, completed by induction with elements concerning availability and efficiency. This way of language design guarantees the completeness of the design; it is given by Steusloff (1977) in some detail (e.g. properties = sets of I/O-values and types; functions = algorithms; structure = communication and data-way tables). Here it is not possible to explain these details but we will discuss the complete language design in the following.

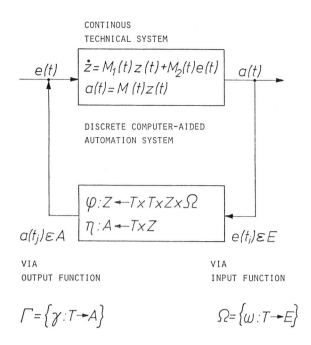

Fig. 3. Features of Technical Systems

Structural Requirements for a Programming Language

The process of programming may be understood as a decision process. The programmer has to make decisions concerning the choice of semantically suitable language elements for a given problem and then must use these language elements in a syntactically correct manner.

From decision theory, it is well known that the decision load depends on the number of decision steps necessary to reach the end of a decision chain, as well as on the number of alternatives at each decision step.

Figure 4 shows the decision load for three different decision models relative to the one-step-model (1) which, in fact, determines one point (x) in Fig. 4. Model (2) represents a hierarchical decision process with a = 2 or a = 3 alternatives for each decision step. Model (3) assumes the number of alternatives increasing by one with each decision step. Model (1) very well describes assembler languages whereas model (3) fits some higher level languages rather well.

From Fig. 4 we can derive that a hierarchical decision model with about 4 steps considerably reduces the decision load. Transferring this to the design of programming languages we conclude that, grouping together language elements for the decription of related process features and then fixing these groups as alternatives in a hierarchical language structure, will reduce programming effort and expenditures.

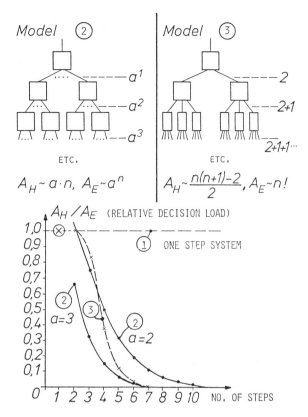

Fig. 4. Decision Models for the Structure of Programming Languages

DESIGN OF AN APPLICATION PROGRAMMING LANGUAGE FOR DISTRIBUTED COMPUTING SYSTEMS

After the above mentioned considerations concerning

- the derivation of the necessary language elements from the principle of duality and
- the minimization of the decision load during the programming process

we have now the tools for designing an application language for distributed computing systems. Instead of giving all the details we will discuss the complete language design by means of Fig. 5, where the structure of model (3) from Fig. 4 is easily recognized.

The presented language consists of two divisions with language elements for the description of the system structure and the functions of the system, the latter ones containing the language elements for the description of system properties. Each of the two divisions comprises three subgroups which, on their part, are again subdivided down to the single language constructions.

It should be pointed out, that the term "language element" in the following means "genuine language element". Thus the non-specific element "CALL name (parameters)" is not a "language element", because the name and the parameters of a CALL don't represent standard and fixed semantics in the context of a programming language. Therefore the notation of a CALL doesn't contain computer-independent and self-documenting information being characteristics for genuine language elements.

All language elements not common in well known and wide-spread programming languages are double-framed in Fig. 5. A dotted frame indicates that in some very modern languages there are some language elements of this type.

Functions and Algorithms Description

This language division starts with the well known language elements for the definition of data and program objects. For distributed computing systems the existence of global objects is essential; these global objects (e.g. global semaphores) are necessary means for system-wide communication and synchronization.

The language elements for formulating algorithms also are well known in all higher level languages, whereas language elements for input/output are usually only available for the input/output from/to standard peripherals and not for process peripheral devices.

Language elements for real-time program flow control and communication control are very new in higher level programming languages. Nevertheless, they are most important to achieve portable real-time automation programs, especially with respect to time-dependent programs, where the use of physical time-units instead of counter-intervals is required.

Resources and Configuration Description

This language division comprises all the structural description elements. These are the description elements for the medium-term unchanged features of each partial computer system in a distributed system structure, the program configuration description elements, and finally, language elements for the description of data ways and communication links.

All these language elements today are not contained in computer programming languages, because their necessity arises only with distributed computer systems, where the structure of a system cannot be predefined due to structural changes during the operation and the lifetime of a system.

There is some discussion about the necessity of an explicit description of the program configuration within a distributed system versus an automatic production of such a configuration by the system itself. At present, we feel that an automatic distribution of programs is not feasible due to the lack of a suitable method for the formal description of program semantics. This would be necessary to determine, for any system state, the optimal assignment of programs and processors. Today only the system designer will be able to perform this task, also and especially for non-regular system states.

Reactions to Non-Regular System Operating States

Most of the language element groups at the third level in Fig. 5 contain elements for the reaction to non-regular system operating states, i.e. for exception handling.

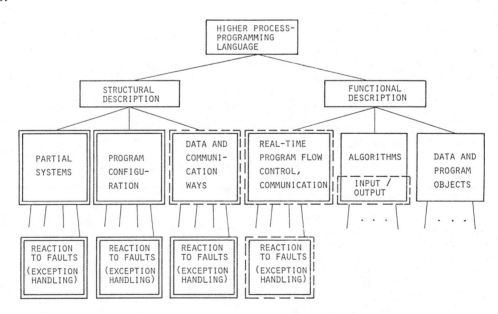

Fig. 5. Basic Structure of an Application Programming Language for Distributed Automation Systems

According to the above mentioned language requirements, all this exception handling is explicitely done by the programmer, who describes transparently the system characteristics for any system operating state. This even may be examined by technical supervision authorities. These language elements should be separated from other language parts to support transparency and, as far as possible, avoid influences from the change of other system features.

REALIZATION OF A SUITABLE PROGRAMMING LANGUAGE FOR DISTRIBUTED COMPUTING SYSTEMS

The realization of an application programming language for distributed computing systems, according to the mentioned design criteria, should start from the analysis of existing languages to find a suitable candidate. If possible, the completely new development of a language should be avoided in order not to re-invent well established language parts, among other things rise to new opinion-directed discussions. Therefore a set of modern real-time programming languages has been examined by means of a value analysis choosing the criteria according to the double-framed language parts of Fig. 5. The method taken for the value analysis is the binal evaluation; i.e., the fulfillment of a criterion yields the value of "1", in the opposite case the value of "0" is given. To determine the final values of utility, all the single values are linearly summed up. This method does not take into account the possible effects of combinations of criteria. It is only allowed, if the criteria are disjunctive and if the candidate with the highest value of utility shows "0" only for those criteria, which are not fulfilled for every other candidate in the analysis.

Figure 6 shows this value analysis which, as usual for every value analysis, contains some subjective opinions of the author. The candidate with the highest value of utility is the language PEARL (Process and Experiment Automation Real-time Language) which therefore is taken as the basis for an application programming language for distributed computing systems. Further on, this language will be called "PEARL for multicomputer systems".

The Programming Language PEARL

The programming language PEARL has been defined and developed by a German group of people from industry and scientific institutions with the support of the German Ministry of Research and Technology (data-processing programs). The algorithmic part is related to ALGOL 68 and PL/I, among others comprising pointers and the definition of compound data structures and types. The program structure is block-oriented; a MODULE is subdivided into TASKs and PROCEDUREs, on the MODULE-level as well as on the block level (Kappatsch and others, 1978).

There are genuine language elements for real-time program flow control and explicit synchronization of TASKs for event-controlled program flow and for the control of parallel activities. Besides the language elements for standard-I/O and formatting, there are special elements for process-I/O. This all is part of the PROBLEM-division of a PEARL-MODULE.

In a second division, the SYSTEM-division, a PEARL-MODULE contains an I/O-data-way description. This is one of the outstanding features of PEARL, giving means for some structural description of a single-computer system by a separate division of a MODULE.

On the basis of these features, PEARL is the most suitable candidate for a language according to the design of Fig. 5. However, PEARL does not contain language elements for the structural description of a distributed computing system and its programs. Furthermore, exception handling has to be programmed within the algorithmic part, even for structural or I/O-faults (double-framed blocks in Fig. 4). To cope with this situation we have added suitable language elements to PEARL.

PEARL Extensions towards "PEARL for Multicomputer Systems"

Before presenting these PEARL extensions, as they are defined by Steusloff (1977), it is necessary to give a statement concerning the situation of PEARL. At present there are two completely defined versions of PEARL, Basic PEARL (Hruschka and others, 1978) and Full PEARL (Kappatsch and others, 1978). The first one has now been standardized in the FRG (DIN 66253, 1979) and has been submitted to ISO for standardization, whereas Full PEARL is in the process of standardization in the FRG. There are at least six implementations in the FRG and one from abroad which are or will be available in 1980 at the latest. Therefore PEARL for multicomputer systems must not change standard PEARL in any way, rather than literally extend PEARL in such a way that a standard PEARL compiler would understand the standard program parts of a program written in PEARL for multicomputer systems. This is achieved, according to the above mentioned structural design requirements, by adding on new language divisions or using the technique of "pragmats", i.e. special comments.

Description of Partial Systems (STATIONS-Division)

This additional division of "PEARL for multicomputer systems" serves for the description of the static structure of a distributed computer system. For each separate computer (processor unit) three groups of components are described, namely the processor, the I/O process peripherals and the functions of the

Genuine language elements for	Application					System Implem.		
	RT-FOR	RTL/2	BASEX	PROCOL	PEARL	JOVIAL	CORAL	Ada
Real-time control	∅	∅	1	1	1	∅	1	1
Synchronization (processes, resources)	∅	∅	∅	1	1	∅	1	1
Process-data-I/O	∅	∅	1	1	1	∅	∅	∅ (1)
Bit processing	∅	1	1	1	1	1	1	1
Global program communication	∅	∅	∅	∅	1	∅	∅	1
Hardware structure	∅	∅	∅	∅	∅	∅	∅	∅
Software structure	∅	∅	∅	∅	∅	∅	∅	∅
Dataway & Communication structure	∅	∅	∅	∅	1	∅	∅	∅
Exception handling	∅	∅	∅	∅	1	1	∅	1
Value of utility	∅	1	3	4	7	2	3	5 (6)

Fig. 6. Value Analysis of Programming Languages

local operating system (Fig. 7). These specifications are required, since, in general, different processors and hardware realizations adapted to the local needs will be present.

portant: STA1PR may indicate a failure of the processor, STA1COM the loss of the communication of the partial system named STA1. By means of these status-codes, the use of function-sharing redundancy is controlled. Status-codes are produced by the error detection devices or programs of each MCS (Fig. 2).

Description of devices follow as a list of properties and register addresses for the description of a prespecified basic structure of device registers; registers not present in the actual device are marked by NONE. Device names must be unambiguous within one overall system; even though basically any name is legal, meaningful names are recommended to increase legibility of programs (e.g. ANIN56∅ = Analog Input, Station No. 56, Device No. ∅).

The partial system description is ended by the description of the operating system functions, present in a partial system, by the corresponding PEARL-keywords. This information among others is used for test purposes by the program generation and loading system and may also serve as generating information for a modular, task-adapted, local operating system.

The static description of the overall distributed computing system results from the collection of all partial system descriptions by the keywords STATIONS and STAEND. Therefore, it also contains all those resources which are available to the overall system as function-sharing redundancy.

```
STATIONS;
NAME            : STA1,56;
PROCTYPE        : IITB;                              processor
WORKSTOR        : 5600000, 5649142;                  description
STATEID         : (STA1PR:H'188'),(STA1COM:H'430');
  :
DEVICE ANIN56∅  : ANIN∅, IN, WORD, FIXED (15),
                  H'F101', H'F102', H'F103',         I/O devices
                  NONE, NONE, H'F104', NONE;         (register
DEVICE ANOUT56∅:  ...                                addresses)
  :
OPSYS           : (ACTIVATE, TERMINATE, SUSPEND,
                  CONTINUE),
                  (AT,ALL,UNTIL,WHEN),               functions of
                  (ENABLE, DISABLE, TRIGGER),        operating
                  NONE, NONE, NONE;                  system
NAME            : STA2,57;
  :
STAEND;
```

Fig. 7. Partial System Description

A partial system description has to begin with the name of the partial system and a station number, followed by the characterization of the type of processor, e.g. by the name of the code generator for this type of processor. The keyword WORKSTOR characterizes the notation for the boundaries of the working storage. The combination of status-identifiers with codes for the operating state (status-codes) is im-

Description of Reactions to Changes in Structure of Distributed Computing Systems

The structure of distributed computing systems is changed by disturbances and failures of processing units, field-units and communication components, as well as by planned extension or reduction of the systems. The handling of disturbances and faults consists of the use of function-sharing redundancy. Describing aids for this purpose, with respect to processing functions (programs) and process data, are displayed in the following by means of examples as extensions of PEARL-MODULES.

Program configuration and reconfiguration.
Language elements for the description of program configuration and reconfiguration, depending on the operating state of the computing system, may be called "load instructions" (Fig. 8). They consist of the specification of a destination computer station, of the so called load priority, of an operating status condition and optional supplements. The load priority determines the importance of a program module relatively to other modules within a partial system (functional priority); this is to be distinguished from the operational priority of individual subprograms. The operating state INITIAL denotes the initial (bootstrap) and normal status, while the logical expressions in parentheses describe non-regular operating states by means of status-identifiers from the STATIONS-division. The optional specification of a starting number for the initial status describes the bootstrap-sequence of a program system. The optional attribute RES[IDENT] causes the bootstrap-loading of programs into redundant partial systems, in order to be available there after failure of the program loading device or to shorten the loss of processing functions after time-critical program reconfigurations.

```
        MODULE MOD1;
        LOAD;
        TO STA1 LDPRIO 5 INITIAL STARTNO1;
        TO STA3 LDPRIO 5 ON(STA1PR.AND..NOT. STA3PR)RES;
        TO STA2 LDPRIO 5 ON(STA1PR.AND. STA3PR);
        SYSTEM;
            :   data-ways, correction of data, description of replacement data
            :
        PROBLEM;
            :   description of algorithms
            :
        MODEND;
```

Fig. 8. LOAD-division

Replacement process-data-ways and replacement values. The language elements for the description of replacement process-data-ways and replacement values after disturbances are arranged in a data-way description (PEARL-SYSTEM-division), in order to separate the description of data transfer and data generation from the processing algorithms (Fig. 9).

The name TEMP of a source of process-data may denote a temperature input, the sensor of which is, in the standard case, connected to device ANIN560 (cf. partial system description), channel 8, an analog input. Alternative connections are supposed to exist to devices ANIN571 and ANIN572 and, if necessary, also from additional (redundant) sensors. For these alternatives, correction algorithms for the adaption of the measured data to different properties of devices and data-ways can be specified. The last alternative denotes a replacement value. In the same line there are also the test conditions for the plausibility test of the measured data obtained.

```
TEMP :  -> ANIN560 * 8
  /*$   -> ANIN571 * 5,
            CORR: TEMP = TEMP/3-16
        -> ANIN572 * 5,
            CORR: CALL ADJUST (TEMP,2)
        -> REP: TEMP = 1300,
            PLAUS: (HI=1600,LO=100,
                DELTA=10)*/;
```

Fig. 9. Data-ways, data-corrections, replacement value

Such a data-way description is processed in the order of its notation, depending on the operating state of the data-ways and the result of the plausibility test. If all data-ways are disturbed or produce non-plausible values, the replacement value is taken as the measurement value. Thus a valid value is always available to the variables of the algorithmic program sections (PEARL-PROBLEM-division), without burdening the processing algorithms with all the exception handling associated with process data I/O.

SUPPORTING SYSTEM PROGRAMS:
THE PROGRAM GENERATION AND
LOADING SYSTEM

The above described parts of a multicomputer real-time programming language require a special program generation and loading system. In general it comprises four functional groups, the compilation of a higher level programming language, the code editing system, the code loading system and a debugging and testing system.

As shown in Fig. 10, source program modules are translated into an intermediate language (IL1) which is written into a library. This intermediate language is assembler-like but not dedicated to a particular instruction set;

i.e., the complex statements of higher level languages are decomposed into a sequence of operations. By this means, the intermediate language serves as a link between higher level languages and the machine-dependent machine level languages, thus avoiding the need of many special compilers in a heterogenous multi-computer system.

The local system-state observers observe all the system-state information generated by the error detection units in the processor stations and communicated through the system by the global communication system. The observers react only if the state-information corresponds to an entry in their specific local status-table.

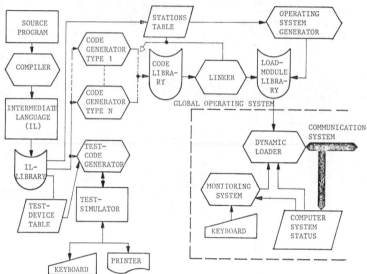

Fig. 10. Program Generation and Loading System for Distributed Real-Time Computer Systems

According to the processor specifications contained in the load part of a PEARL-MODULE, one or more code generators are selected, each of them producing a relocatable machine code version of the same source program. The relocatable codes are linked together with run-time subroutines and operating system functions; the linked load-modules are then stored on disk. By this, program reconfiguration in accordance to the load parts of a program system and the state of the computer system may be executed very fast just by down-line loading of ready-to-run load modules. The dynamic loader (Steusloff, 1977) will execute this task, comprising a decentralized, distributed local system-state observer and a centralized global loader.

The local system-state observers act independently upon status-tables, each of them containing all the locally relevant information, condensed from the LOAD-parts of all the PEARL-MODULES of an entire automation system. This information consists of status-expressions ("INITIAL" or "ON (status-expression)"; cf. Fig. 8), the related MODULE-names and the actions scheduled for the case when a status-expression becomes true. Actions can be

- start of a pre-loaded MODULE (attribute RESIDENT in the related LOAD-statement) or
- load a MODULE from the load-module library via the global loader and the global communication system, then start the loaded MODULE.

Actions comprise global checks to maintain an unambiguous system-state concerning the activities of MODULES; e.g., it must be sure that the same MODULE is activated only in one of the processor stations, although it may have been loaded into several processor stations of the system.

These problems are discussed in Heger and others (1979) and some of them are still research work.

A HIERARCHY OF LANGUAGE LEVELS

The program generation and loading system shows several language levels from the PEARL-level down to the machine instruction level (load modules). In Fig. 11 these levels are represented, along with a level of very high level languages (VHLL), which are dedicated to special problems. Therefore they are called problem-oriented languages (POLs), whereas higher level languages like PEARL or FORTRAN are universal and operative languages, i.e. not problem-oriented. Intermediate languages usually are machine-independent and either assembler-like or operator-languages. They can be interpreted, mapped to a machine-dependent assembler or directly translated to machine-code. The special class of system implementation languages (SPL), like CORAL or Ada, can be located between the HLL and the IL in Fig. 11. Among others, they can be used for the implementation of the HLL and VHLL.

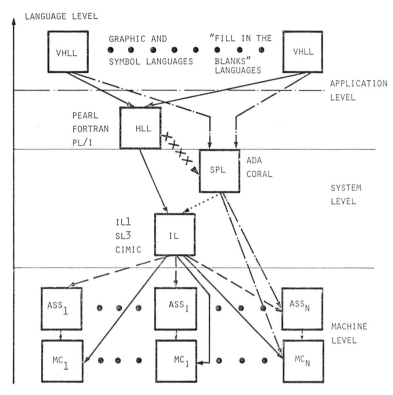

Fig. 11. Language Levels and Interdependencies

The strong lines in Fig. 10 indicate the use of a HLL as an intermediate language for VHLL, employing then an IL to serve heterogeneous computing systems. It may be convenient to interpose an assembler language for easier testing and software portability (dashed lines). The dashed-dotted lines show the way via a SPL, which generally will produce machine-level programs but may also use the IL-way (dotted line). The x-line is of some importance: In future, the SPL will probably be used to implement HLL as well as serving for the production of machine-level programs.

There seems to be a competition between the SPL and the HLL. But looking at the different levels of both language classes, one can easily see that they complement one another, if we agree that for the coming years it will be necessary to have both the HLL and the VHLL. For the sake of portability, transparency and self-documentation of automation system application programs, the application programmer will need HLL, whereas the system programmer - with the same requirements - needs a machine-independent language for mapping automation system functions to a machine level. Bootstrapping the language hierarchy of a system, comprising a SPL, it may be efficient to compile a HLL into this SPL, using the SPL as an intermediate language and, on the other hand, implement the HLL-compiler by use of the SPL. Thus both language types have their fields of application and will help to improve the efficiency of software production.

REFERENCES

DIN 66253 (1979). Programmiersprache PEARL, Basic PEARL. Deutsches Institut für Normung (DIN), DIN 66253, Teil 1.

Heger, D., H. Steusloff, and M. Syrbe (1979). Echtzeitrechnersystem mit verteilten Mikroprozessoren. Bundesministerium für Forschung und Technologie. Forschungsbericht DV 79-01 Datenverarbeitung. (Real-time Computer System with Distributed Microprocessors. Federal Ministry of Research and Technology (FRG). Report DV 79-01 Data Processing (in German)). 364 pp.

Hruschka, P. et al. (1977). Basic PEARL Language Description. Kernforschungszentrum Karlsruhe (FRG). Report KFK-PDV 120, 242 pp.

Kalman, R. E., P. L. Falb, and M. A. Arbib (1969). Topics in Mathematical System Theory. McGraw-Hill, New York.

Kappatsch, A. et al. (1977). Full PEARL Language Description. Kernforschungszentrum Karlsruhe (FRG). Report KFK-PDV 130, 462 pp.

Knight, K. E. (1966). Changes in Computer Performance. Datamation 9, pp. 41-54.

Leinemann, K., and U. Schumann (1973). Eine Wortsprache zur Beschreibung technischer Objekte. (A Language for the Description of Technical Objects.) Lecture Notes in Computer Science, Vol. I. Springer-Verlag, Heidelberg. pp. 465-473.

Steusloff, H. (1977). Zur Programmierung von räumlich verteilten, dezentralen Prozeßrechnersystemen. (Programming Spatially Distributed, Decentralized Process Computing Systems.) Dissertation. Fakultät für Informatik der Universität (TH) Karlsruhe (FRG). 180 pp.

Steusloff, H. (1979). Structures of Automatic Control Systems and the Consequences for Programming Languages. PA Process Automation 1, R. Oldenbourg Verlag, München. pp. 3-11.

Syrbe, M. (1978). Basic Principles of Advanced Process Control System Structures and a Realization with Optical Fibres-Coupled Distributed Microcomputers. Preprints of the 7th Triennial World Congress, IFAC, Helsinki. Pergamon Press, New York. pp. 393-401.

DISCUSSION

Buchner: The class that a particular function belongs to might depend on the state of the controller, communication topology, and the task locations. You seemed to have solved that; at least you indicated a solution in the PEARL language by evaluating logical expressions for loading a particular module into a particular processor. Do you consider that, in many cases, the alternatives to where the task may be located because of the differ-ences in the state of the process and the controller, may be so complex as to really preclude an a priori determination of these factors?

Steusloff: I agree, that may be very complicated. But when designing such a software system, the designer has to think about these very complex situations and the LOAD statements give a means for writing down the result of this thinking. No language itself solves your problems; the system designer has to solve them, but having done so, he needs language elements to put the solution down. Does that answer your question?

Buckner: Yes, it does.

Steusloff: Let me make one additional point. The interdependence of all these states may be very complicated and it, at least, should be possible to find out whether or not they are contradictory. We have added to our compiler a part which proves that those status expressions in the LOAD statements of a module are not logically contradictory. But the system designer has to know what states are relevant and where system resources can be employed, under what circumstances, and so on.

Miller: Could you further elaborate on what you see as the requirements for concurrency mechanisms, or control of such, in a language considering two subcategories: one, in terms of interprocess communication and coordination; for example, of tests wanting synchronization; and, second, concurrent data base updating operations?

Steusloff: For the first question, I would have to go rather deep into PEARL. PEARL has a lot of language elements which support concurrency of tasks, scheduling of tasks, and synchronization of tasks and access to data. The problem for us is to expand these features to a distributed system. The concurrency of tasks and these other problems are solved by local operating systems along with the assistance of a higher level global operating system. I can't give you all of the details here but some of them are in the paper. The problem of access to a distributed data base at present is solved in a very simple manner due to the lack of better realtime-oriented solutions. Let me give you an example: We have some very important global data base parts. These are the tables containing the location of global objects, which can be accessed from all over the system. All these tables have to contain exactly the same information in all stations at the same time, and for all times. This is a distributed data base but it is distributed by copies, which makes problems worse. We have done the synchronization in a very simple way; we send a telegram all over the ring that says no one has the right to access these tables. All the stations have to respond that they have finished with all access to this table which was in progress. Then the tables are changed and another telegram is communicated all over the system that says the data base is now freely available. It is very simple and we are sure there are some better means being studied in data base research.

Miller: Are those mechanisms explicitly defined in your language? In other words, do underlying mechanisms allow all of these good things to happen?

Steusloff: For the example I gave, the underlying mechanisms do all of the job because the access to global data is routed via these tables and programmers only employ the GLOBAL attribute in their programs for systemwide communications. In addition, there are further explicit synchronization elements in the PEARL language.

Miller: Could you discuss the efficiency versus portability tradeoffs in the usage of your intermediate language concept?

Steusloff: Well I'm afraid we would have to start with the efficiency problem of higher level languages, and I don't think we can discuss it here in detail. But we have solved some efficiency problems with PEARL by giving some rules of using PEARL. That worked quite well. These rules are different for different functions. The restrictions are very hard for the indispensable functions and they grow weaker for the parameterizable functions or the dispensable functions. Usually the problems of efficiency with the latter class of functions are lower than the problems of efficiency with the class of indispensable functions. That's how we did it, but it is a very pragmatic way of solving the problem. But I can't give you comparative numbers now. Concerning the intermediate language, we use an operator language which does not really affect the efficiency of the running programs.

Miller: You mentioned which procedures you have to follow when you change tables. Could you give a figure on the worst case reaction time under such conditions?

Steusloff: It depends very much on the communications system and I can only speak for our system. The average communication time for ringwide communication is about 10 milliseconds. That's because we have an active coupling of the stations; all the telegrams have to go through all the stations and there is a certain delay in each of the stations. This results in about 10 milliseconds for a ring with about 32 stations and under the condition that the stations forward the telegrams immediately and do not send other telegrams in between. As I told you, there are three telegrams necessary, so in the worst case it may be, very roughly, 30 milliseconds.

Müller: But then you have to change the tables, and that takes time.

Steusloff: That is one of the telegrams: One telegram stops all access to the tables, one telegram updates the tables, and one telegram makes the tables available again. One telegram contains 64 words of 16 bits each.

A DISTRIBUTED CONTROL SYSTEM— CONCEPT AND ARCHITECTURE

S. Yoshii and S. Kuroda

Hokushin Electric Works, Ltd., Tokyo, Japan

Abstract   This paper presents a distributed control system concept and architecture which has been implemented in 900/TX, a microprocessor-based instrumentation system developed and marketed by Hokushin Electric Works, Ltd.
The hierarchical multilevel systems approach is made, taking into consideration the requirements of plant operation with emphasis on man-machine interface and intersubsystem communication.

1. INTRODUCTION

Improvements in plant operation in terms of process efficiency, reduced pollution and improved working environment can be achieved by integrating interrelated process units and by introducing centralized and sophisticated control and supervision equipment. A trend in order to achieve a better overview of total plant operations has been the provision of centralized control rooms with the aid of introduction of miniature electronic conventional instrumentation. Computer technology also began to be incorporated in the centralized control philosophy.

The centralization and sophistication, however, lead to an increase in size and complexity of the instrumentation system, thus causing the following drawbacks, which in turn have given impetus to decentralized scheme of instrumentation;

- Increase in cost to gather necessary information, due to geographical distribution;

- Reduced reliability and extended scale of system destruction by a single failure, causing loss of system availability;

- Difficulty in designing and maintaining the system; and

- Degradation in response time.

The first successful approach to decentralization was the functional decentralization based on hierarchical multilevel system approach in which the total system is vertically decomposed into smaller manageable subsystems, normally having classical type of feedback control structure on the lowest level, while the higher levels involve more sophisticated functions. This approach resolved at least the above mentioned third and fourth issues.

To make an effective approach to the first and second issues of the drawbacks, however, we have had to wait until physical decomposition became economically feasible. The most influential technological impact that enabled physical decomposition has been the cost reduction in installation works by line sharing technique and the remarkable advances and cost reduction of microprocessors.

Thus, it may be said that the recent technological innovative advances together with the available systems approaches have enabled the development of a viable distributed control and supervision system which resolves most of the aforesaid disadvantages brought up by the centralization and sophistication.

On the other hand, the introduction of CRT displays for industrial application has given an evolutional impact on improvements of man-machine interfaces. Combined with microprocessor and communication technology, the CRT display has brought powerful means to process available information in order to show it to the operator in the most comprehensive form.

Following discussions will describe a design concept to build up a distributed control system architecture with now available resources and the discussion will extend to the description of 900/TX Distributed Control System designed under this concept.

2. DESIGN CONCEPT

Centralization and sophistication are an irresistable process which has been occurring for decades to achieve improvements in plant operation. Decentralization and simplification are not a counterprocess to this at least in the field of instrumentation. They are a movement to give methodologies with which centralization and sophistication can be achieved without causing the drawbacks mentioned in Chapter 1.

In this sense hierarchical multilevel systems approach is still the most powerful conceptual and practical tool to construct a large and complex control system.

The idea that direct control functions are placed on the first level while more sophisticated functions are implemented on the higher levels seems still persuasive.

However, the situation has been changed since the idea was first presented.

The technological innovation in the field of solid state devices has enabled to provide a powerful but economical approach to the design of computer-oriented systems even on the first layer of the hierarchy.

Also, the advances in communication, especially data communication offer flexible and economical means for subsystem links needed for co-ordination at higher levels.

Thus, in the field of instrumentation, the hierarchical multi-level approach is not only conceptual and functional but physically viable in its almost ideal form.

Although the decision based on a single variable could be distributed to the extent that the decision maker is located at the control point of the plant, there is practically no advantage over that scheme in which the decision maker is located in the control room, since the decision maker requires several decision parameters which are conveniently set up in the control room. Therefore, the distribution in the view of decision maker may be limited at the most to the control room equipment. By extending this argument, it could well be said that the distribution of the system made up by computer-oriented approach will stop at the level on which the present analog control scheme stands.

Even with the latest technologies, however, it is still not feasible to make a computer-oriented approach to the decomposition of the first level into the subsystem with only basic loop control function such as conventional PID function.

It is therefore still practical to utilize the capacity of the second layer to realize such functions, leaving on the first layer the functions of basic manual control and interaction with the process, or to use conventional analog controllers, as first layer subsystem, with the function of communication with the higher levels.

On the second layer, microprocessors play an essential role. It is now economically feasible to design a self-closed control system with a built-in microprocessor to cover all the required functions to operate one or several process units. The size or capability of the system may depend upon the targeted application area.

The third layer is reasonably constituted of mini-computer level subsystems and centralized operator interface subsystems. Data highways have already established their position as the most flexible and economical means for communication link between the second and the third layers.

Another important issue in control and supervision system design is man-machine interfaces. To make a unified approach with a universal instrumentation system to satisfy varieties of plant operation, it is necessary to have versatile man-machine interfaces. For example, relatively small scale systems may be economically instrumented with individual loop interfaces which have long been used in analog instrumentation. In such transitional status of plant operation as start-up, shut-down or contingency, it may also be justified to have panel boards mounted with man-machine interfaces with which simultaneous, loop-by-loop operation is carried out by several operators. On the other hand, relatively large and stable processes may be manipulated through centralized operators console by use of shared displays, control-by-exception and other techniques.

Man-machine interfaces based on the panel boards of conventional type have been accepted by the operator. There are, however, needs and a room for innovative consideration to get rid of the mess of interconnection wirings inside the panel boards and to achive ease of design modification after actual installation. Line sharing may be effectively used to tackle with this problem.

Although the idea of plant operation with centralized operators console is relatively new and the standard practice in designing the console is yet to be established, one of the design targets should be to maintain as much as possible such positive features of the conventional panel boards as individual and parallel access to any variable, one-to-one correspondence between operational actions and operational means, system overview by the display patterns on the panel boards, displays with scales, bar graphs, enginnering units, tag numbers, trends, etc. Another target should be operational simplicity. For example, the shared display and operation is accepted by the operator only when the system is so designed that shift from one page to another can be made by simple and minimum actions.

The last but not least issue to be mentioned in the design of distributed control system is software. Most distributed control systems are designed as a partial replacement of conventional analog controllers.

They involve a mass of persons who are trained and accustomed only to conventional analog instrumentation. Software, as well as the other elements of the system should be so conceived and structured that, with minimum period of training, it is accepted by system designers, process engineers, operators and maintenance engineers who have been familiar with conventional analog instrumentation.

Techniques centered around modularization of the program in simple functional blocks may provide an effective tool to deal with this issue.

3. 900/TX SYSTEM DESCRIPTION

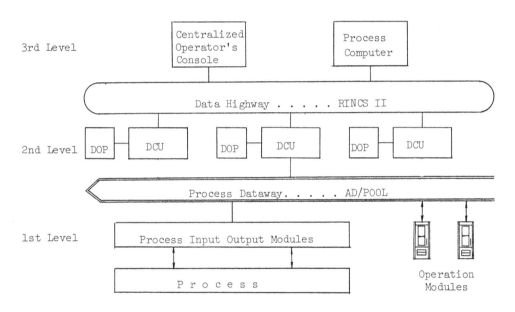

Fig. 1 Simplified Diagram of System Configuration

3.1 System Configuration

900/TX is decomposed into three levels of subsystems as shown in Fig. 1.

The first level provides interfaces with the process and individual loop-by-loop machine interfaces. The process interfacing is carried out through process input-output modules, each of which processes a single analog variable or 16 on-off variables. The same operation as with conventional analog controllers can be made on the Operation Modules, each of which is housed in an independent case. Communication between the Operation Module and the process input-output modules are made via AD/POOL, a process data way.

The second level gives the integrated functions required for the control of one or several process units by implementing a combined function of continuous reguratory control, sequential control and advanced functions such as multivariable control. A subsystem on the second level may have a locally centralized operators console, either dedicated to it or shared with other subsystems.

On the third layer, process computers and centralized operators consoles are located. With a store-and-forward unit, computers of other make can be a member of the third level subsystems. Second level subsystems are linked to these subsystems via RINCS II data highway.

3.2 AD/POOL and First Level System Configuration

3.2.1. AD/POOL

AD/POOL is a process dataway which provides information exchange paths between the modules connected to it. It carries analog and discrete signals cyclically on time sharing basis as shown in Fig.2.

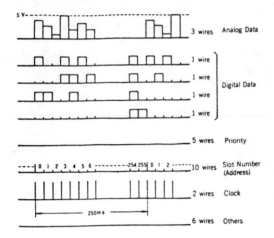

Fig. 2 Signals on AD/POOL

AD/POOL consists of a 30-core cable and a master unit which generates 256 "time slots" cyclically at the period of 250 ms. An information source transmits data onto AD/POOL only when the time slot appears of which slot number coincides with the address assigned to the information source. More than one information source may be assigned the same AD/POOL address. Conflicts between the sources with the same AD/POOL address are resolved by priority control. Each information source is given a priority level and in case that more than one source try to make an access to AD/POOL, only the source with the highest priority level is allowed to transmit data onto AD/POOL. Five priority levels are provided. An information sink on AD/POOL picks up data on the time slot of which number is the same as the AD/POOL address assigned to the sink. More than one sink may be assigned the same AD/POOL address and pick up data on AD/POOL at the same time.

Analog data are transmitted in a 4 to 20 mA current signal from an information source, converted into a voltage signal by the master unit and read by sinks. A source transmits 4-bit discrete data onto a time slot, requiring for example, four time slots to send 16-bit data.

By the configuration mentioned above, the source to sink correspondence can be made or modified simply by assigning or changing the AD/POOL address of the corresponding modules on AD/POOL, thus enabling to proceed the system design and the panel wirings independently. Even after installation the information paths between modules on AD/POOL can be set up or changed without difficulty.

3.2.2 Modules on AD/POOL and Operating Principle

The following are provided as the standard modules that can be linked via AD/POOL. However, any module with the standard AD/POOL interface may be connected to AD/POOL.

- <u>DCU</u>
  A second layer microprocessor-based controller. Can make an access to any time slot on AD/POOL.

- <u>Operation Module</u>
  A panel board mountable module for loop variable display and loop operation. Includes the circuitry for manual control. The standard PID control action can be performed here as option. Four AD/POOL addresses are assigned to this module, each for setpoint, process variable, control output, and control output answerback.

- <u>PV Module</u>
  A module to convert an analog process variable into a standard AD/POOL signal. Mounted in a rack case and assigned one AD/POOL address.

- <u>MV Module</u>
  A module to convert a control output from DCU or an Operation Module into a signal suited for the corresponding actuator. Mounted in a rack case and assigned two AD/POOL addresses, one for the control output and the other for the control output answer-back.

- <u>Contact Input Module, Contact Output Module, Pulse Input Module</u>
  Modules for contact signal input, contact signal output, and pulse train signal input, respectively.

- <u>Sample and Hold Module</u>
  By this module, any signal on AD/POOL can be accessed, picked up, and converted into continuous analog signal for indication, trend recording, etc.

To show the operating principle, let us take an Operation Module linked with a PV Module and an MV Module. For a correct data transmission, the AD/POOL address for the process variable of the Operation Module and the AD/POOL address of the PV module must be the same. Also, the addresses for the control output and its answer-back of the Operation Module must coincide with those of the MV Module, respectively. In this configuration, the process variable transmitted onto AD/POOL from the PV module is picked up by Operation Module for indication. If the Operation Module is equipped with PID calculation circuitry, the process variable is compared with the setpoint which is given either from DCU or from the setpoint thumb wheel on the front panel of the module. PID calculation is performed on the deviation obtained by the comparison and the result is transmitted to the corresponding MV Module via AD/POOL. The control output to the process from the MV Module can be read back to the Operation Module for checking and other use.

3.3 DCS (Direct Control Subsystem)

DCS is a microprocessor-based controller for integrated control of one or several process units. It consists of DCU - processer unit, DOP - local operators console, and modules on AD/POOL. DCU has an access to any variable on AD/POOL and can transmit an output onto any time slot on AD/POOL.

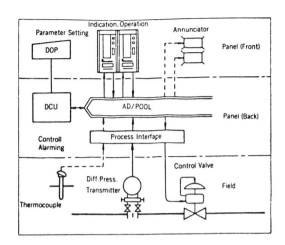

Fig. 4 Cascade Control Loop Instrumented with 900/TX

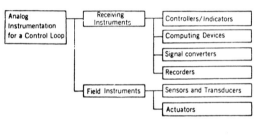

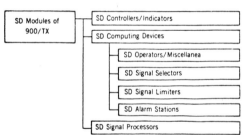

Fig. 3 Analog Instruments and SD Modules

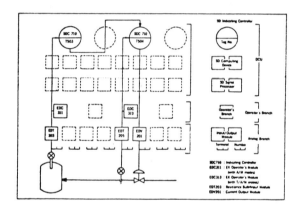

Fig. 5 Loop Description Sheet

The program for continuous process control in DCU is built up by "soft-wiring" modular program blocks named SD-module. SD modules have been designed in such a way as persons accustomed to analog instrumentation can easily get familiar with them. As shown in Fig. 3, almost complete one-to-one correspondence between conventional analog receiving instruments and SD modules is maintained.

An example of a cascade control loop instrumented with 900/TX is shown in Fig. 4. "Soft-wiring" between SD modules is made by drawing wiring diagrams on Loop Description Sheets (Fig. 5), punching the information from the Loop Description Sheets onto cards, compiling them on a host computer, and finally loading the object code onto DCU. The program for sequential control in DCU is written in a sequence decision table language. The control scheme described in sequence decision tables is translated by a host computer into a program code loadable onto DCU.

3.4 Man-Machine Interface and COS (Central Operators Subsystem)

3.4.1. Man-Machine Interface

As mentioned in Section 3.3, 900/TX provides an analog-like, individual loop man-machine interface which looks, to the operator, the same as EK instruments, Hokushin's analog controller series. Only difference between EK instruments and 900/TX Operation Modules is that with 900/TX, the calculation is usually made by a microprocesser inside DCU while with EK controller, it is performed by individually built-in analog circuitry.

Another form of man-machine interface - centralized man-machine interface - is provided through color CRT based operators consoles. These two types of man-machine interfaces can be used, conbined or alone, in a form best suited for the type of application.

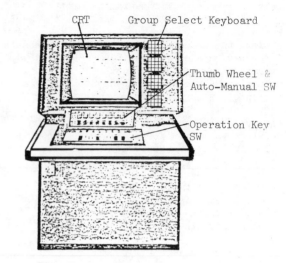

Fig. 7 CRT Display and Keyboards of COS

3.4.2. COS (Central Operators Subsystem)

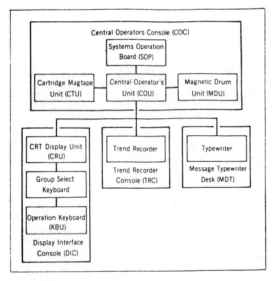

Fig. 6 Simplified Diagram of COS Configuration

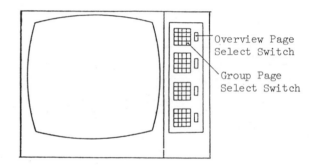

Fig. 8 Group Select Keyboard

Making full use of intelligence obtained by microprocessors, 900/TX provides the functions required for centralized plant operation. Configured as shown in Fig. 6, it allows process information to be shown in three basic display formats.

The overview display prensents a macro-view of a process area or plant to the operator. Up to four pages of overview display are available with each page having up to 128 bar graphs representing process variables of 0 to 100 percent or deviations from setpoint or target value of plus or minus 19 percent. A page is called up by pressing the corresponding overview select button on the Group Select Keyboard located on the righthand side of the CRT (Fig. 7, Fig. 8)

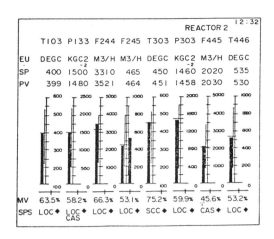

Fig. 9 Group Display

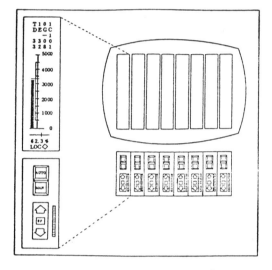

Fig. 10 Thumb Wheels and Auto-Manual Switches

Eight thumb wheels and auto-manual switches are mounted on the Operation Keyboard located beneath the CRT. (Fig. 10) Each loop displayed on the CRT can be manipulated using the thumb wheel and auto-manual switch right below itself. When the auto-manual switch is in the manual position, thumb wheel operation changes control output. Otherwise, it causes setpoint change.

Loops may be formatted on a page in any combination so that loops in given area or processes may be shown together. Loops may be displayed on more than one page.

Up to 64 pages of group display are provided, with 16 pages grouped in a page of overview display.

With this configuration, the operator has virtually 64 panel boards with each mounted with eight controllers and can shift from one panel board to another by simply pressing button. Furthermore, a controller can appear on more than one panel board.

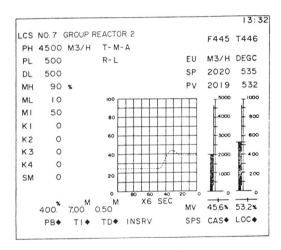

Fig. 11 Loop Display

A group display can be called up by depressing the corresponding group key on the Group Select Keyboard. Up to eight loops appear with the group name on a group display page. (Fig. 9) For each loop, tag number, engineering unit, and loop status are displayed. Also displayed are process variable, permissible ranges of process variable and control output with a vertical scale. A horizontal bar index represents the setpoint value. Setpoint, process variable, and control output are also indicated in numerial digits. Control output is indicated on a horizontal scale and at the same time, if the loop is in the console manual mode, it is displayed on the same vertical scale on which process variable is shown. (F 245 in Fig. 9)

Beneath the CRT, there are also located eight keys named Loop Select Key, each of which corresponds to the loop displayed above it. By pressing a Loop Select Key, the corresponding loop display is called up. As shown in Fig. 11, the loop variables and parameters are indicated together with the trend graph. An extra loop can be called up on the same page by keyboard operation. The loop display may be used for parameter tuning and loop maintenance. Control parameters, setpoint, and control output may also be manipulated through thumb wheel operation. In addition to these three basic displays, multi-trend pages and semi-graphic display pages are provided. Also, pages for maintenance and system modification are included.

3.5 RINCS II and Intersubsystem Communication

RINCS II is a data highway to link the second and third layer subsystems. Physically, it is a loop of twisted-pair cable running between the subsystems. Information is serially and unidirectionally transmitted in a block of up to 124 16-bit words at 200 K bits/second. Up to 32 subsystems can be connected to RINCS II. A subsystem can make a line request at one of two priority levels. If more than one subsystem request line usage at the same priority level, the subsystem located on down stream next to the subsystem at present holding the line control will get the priority.

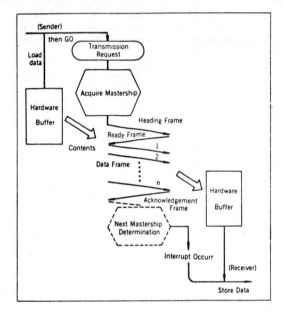

Fig. 12 Procedure for Information Exchange

A 16-bit word to be transmitted through RINCS II is assembled in a 35-bit frame. A frame contains four bits of preamble for synchronization, and five bits of error check codes in addition to the body of the transmitted information. Fig. 12 shows an example of procedures for information exchange through RINCS II.

Protocols for data base transfer and down line program loading between the second and third layer subsystems have been established. For each second layer subsystem, a built-in hardware is provided which allows the third layer subsystems "wake up" and start up the second layer subsystems.

4. FUTURE DEVELOPMENT

The concepts and realization presented in this paper have been well accepted through various applications. However, the form of implementation is yet to be subject to change according to the progress of environments.

Studies and advances in the field of plant operation will lead to improvements of scheme of man-machine interfaces. Continued technological innovation will most likely bring a great step toward further improved system architecture.

A few of the developments which we foresee are:

- Further distribution of decision, eventually to the limit that we mentioned;

- Further unified approach to control room instrument hardware design based on more extended use of microprocessors and supported by sophistication of field instruments;

- Application of the shared line technique to the area of field cabling based on further advances in communication stimulated by optic fiber technology; and

- A more extensive use of CRT display with built-in microprocessor intelligence.

The experience obtained through the conception, design and application of 900/TX will provide a sound basis for the developments of instrumentation in this direction.

5. ACKNOWLEDGEMENT

The authors wish to emphasize that many people played essential roles in developing and implementing 900/TX and that this paper presents the results of their combined efforts.

6. REFERENCES

1) I. Lefkowitz : Multilevel Approach Applied to Control System Design, ASME Trans., 88 (June 1966)

2) M.D. Mesarovic : Multilevel Systems and Concepts in Process Control, Proc. of IEEE, 58-1 (1970)

3) J.D. Schoeffler et al : Distributed Computer Intelligence for Data Acquisition and Control, IEEE Trans. on Nuclear Science, NS-23-1, (February, 1976)

4) J.P. Jansen et al : ICOSS : An Integrated Approach to Process Automation, IFAC/IFIP Internatinal Conference on Digital Computer Applications to Process Control, (June 1977)

5) S.Yoshii et al : Distributed Systems in Process Control, Journal of the Society of Instrument and Control Engineers, (March 1979)

DISCUSSION

Anders: Regarding the justification for decentralization being that it makes the problem simpler, that's at least the second time we have heard that today. Does it really make it simple? Decentralization forces me to decompose the problem, but then good software design on a central system says to break the problem down so that you can handle it. I would go to decentralization more for control objectives, such as performance. Central systems provide management information, save me time, and that kind of thing, and that is always there. But, decentralization would be motivated more from a control purpose. Maybe you could comment a little bit more about how it makes the problem simpler.

Yoshii: I think you are mentioning functional decomposition. If you think about physical decomposition, the situation may change. By physical distribution, you can get high reliability, graceful degradation, and so on. If you limit it to a functional distribution, then, for example, the multilevel systems approach has already solved that problem.

Anders: Those are other benefits, but in terms of simplifying the overall design problem or even writing the software, decentralization introduces other problems; namely, synchronization, copying the data base around was a very simple three-step process, but I needed an optical fiber link in order to be able to do that in the first place, and that is not simple. So I really question the simplicity of design.

Yoshii: If you decompose in the wrong place, that would cause a problem; but this type of decomposition doesn't cause any complexity or problems at all, I think.

Garcia: I have a concern about this architecture: It appears to me that you are retaining a feature of the centralized system, which is the fact that everything has to go through the AD/POOL dataway and that becomes, in a way, a bottleneck because, if that is not up, then all your control functions are down.

Yoshii: We have a duplicated structure for the AD/POOL. We have two AD/POOL's and those two operate alternately every 250 milliseconds. If an AD/POOL fails, the other one takes over. Then, it operates all of the time, not every 250 milliseconds. With this structure, the AD/POOL has very high reliability.

Kuroda: Actually, the AD/POOL is a 30-core cable with a station; for increased reliability, this 30-core cable is duplicated, as are the stations, as Yoshii just explained, to maintain the security of this communication rate.

Garcia: But one feature that I don't quite understand the reason for is, if you have some control function, let's say a simple PID function, that you could accomplish remotely down where the actuator is, you still force the signal to come through the AD/POOL and back out to the field. You do not provide the capability of doing that remotely with a more loose connection to the central system. Is that correct?

Yoshii: This scheme doesn't physically go that far; in this scheme you are still limited to control room devices.

Lalive: I perhaps can try to give an answer to the former question: What are really the benefits of having a distributed system? I think you are right that you should have a good modularity of your system, even in centralized systems, but what you gain with distribution is that, if you do everything together in one centralized system, you need a very complex operating system, which nobody really can understand. These complex operating systems are omitted if you distribute your system, or at least an important part of them is omitted. That is where you get the profit of distributed systems.

May I add a question? It refers to the operator's interface. It seems to me that you have quite a lot of keys, where others prefer to have a light pen, joy stick, a menu, or something like that, on the screen itself. What was the reason for your solution, which is relatively hardwired? You have a fixed, limited number of keys, and it doesn't seem to be too flexible.

Yoshii: That kind of flexibility of operation should be achieved for this process computer. The central operator's console is just the replacement of conventional analog panel boards. As a result, we limited the function of the COS to that function, plus some higher level in operation. Flexibility, in your sense, should be achieved on the process computer level. This is our philosophy.

Lalive: I don't think that answered my question. I mean you have eight hardwired keys you can press because your picture has eight histograms. If you want to have seven or six histograms in order to have more resolution, you cannot do it because you have a hardwired system of eight columns.

Yoshii: I don't understand what you mean by hardwire. Do you mean the correspondence between the functions and the switches is hardwired?

Lalive: I mean the correspondence between the position of the key and the figure on the CRT.

Yoshii: Yes, the correspondence between the slots on the screen and the keyswitches is hardwired. However, the arrangements between the slots and the loops to be displayed are configurable.

Humphrey: I have a question about the AD/POOL, Figure 2 of your paper. First, what kind of distance is the AD/POOL expected to span and, secondly, I find this a very interesting distribution of signals in the AD/POOL. It's the first one in my experience in which there are only four lines for digital data, but ten for addressing data, and then three additional lines for analog data. I would just be curious to hear a little bit of your rationale for such a distribution.

Yoshii: For the first question, the AD/POOL can extend up to 200 meters; so it is limited to a control room. But in real applications, we have some cases where the AD/POOL goes out of the control room to the field.

For the second question, originally, we tried to use only one pair of wires to transmit all of these signals; however, after experimentation, we reached this configuration because the address is the most important thing to detect or locate a failure of modules on AD/POOL. To realize this function, we need as much redundancy as possible for the address lines. That is one reason. We also have three wires for analog data. This is also for use in checking. Any module on the AD/POOL transmits data as a 4 to 20 milliamp signal and this signal goes to the master unit. At the master unit, this signal is converted into a voltage signal. This voltage signal is distributed to all the modules on the AD/POOL analog lines. Then the module that sent the signal can check whether that data was transmitted correctly or not by checking the returned answer. So we need three wires for analog signals.

Korowitz: Do I understand that for linking the SD modules together you need the presence of a large machine to compile it?

Yoshii: To generate the whole system we need a large machine. However, to modify a small portion of the system, we do not need such a machine.

Gellie: I think this is an interesting system you have described, because, the way I understand it, your AD/POOL highway is somewhat different from those most of us are used to seeing. It is somewhere between the 4 to 20 milliamp field wiring and, where we would put a process dataway which is communicating between remote controllers, you seemed to have managed to get this interface nearer the field. It's interesting, but I can't say that I understand the reasons for it. Can you explain what advantages you have? It doesn't seem to map onto what I would call standard distributed processing system. You don't seem to have remote stations, like Garcia said, and you have this unusual highway, like Rusty Humphrey just said.

Yoshii: We have two reasons for introducing an AD/POOL. One is that it saves wiring inside the panel board. Usually, you have many wires since you need one-to-one correspondence between field instruments and panel instruments. So inside the panel, we have very many wires to manufacture. Another reason is that we want to proceed with panel design and panel manufacturing in parallel. Usually you have to design the panel before wiring it. However, with this provision, you can proceed with completely different work concurrently, because wiring can be made by simply specifying AD/POOL addresses.

Gellie: One of the most basic economic advantages of distributed control with serial communication, is that you save a lot of cable costs; but certainly a 30-signal cable is not a cheap cable to be laying about. What's the longest AD/POOL highway that you have installed in a plant?

Yoshii: I'm not sure. Probably we have several applications which use longer than 100-meter AD/POOL cables. Some of them use AD/POOL cables which go beyond the limits of the control room and into the process field.

Johnson: I also was concerned with the reliability of the bus system, not only the AD/POOL but also the ring. You said that you had redundancy of the AD/POOL. How do you handle fault detection on that bus?

Yoshii: We have many provisions for detecting errors in the AD/POOL. For example, we are using a 4 to 20 milliamp signal for transmitting an analog signal; so if two sources on the AD/POOL try to transmit data, they will put current signals on the AD/POOL and these two currents will be added so that it will exceed the limit of 20 milliamps.

Discussion

This is one way of detecting transmission errors. There are many other kinds of detection methods such as this that are used.

Johnson: Is this performed in software or hardware?

Yoshii: Mainly by hardware, but some errors can be handled by the DCU through software.

Buchner: I have another question about the overall architecture of your system. You indicated three levels and, amongst the functions of your second level--the DCS level-- you indicate direct control functions, multivariable control, and continuous time regulatory control. That would imply, if you are going to be using those DCU DOP units, not only is there a significant amount of analog information present on the process dataway, but that the data is also tying up the data highway. So you have information of a significant nature (i.e. very, very low level signals) appearing both on AD/POOL and on the data highway. Am I correct and can you comment on that?

Yoshii: The signal may go up through the DCU and data highway to the central operator's console. However, essential functions are handled in the DCU. So, in this sense, second level functions do not depend on the third level subsystems. Very low level signals do not necessarily appear on the data highway.

Buchner: I understand, but it seems that up to a certain point you are duplicating the functions, at least of your process dataway, in your data highway. In other words, you have two rings, two highways, (the data highway and the AD/POOL), and the data highway duplicates much of the information on the AD/POOL, information that requires significant bandwidth and significant data rates.

Kuroda: This is rather different because the DCU's have certain functions and without the centralized operator's console, with only the DCU and process input-output modules including the process dataway, this system could be a standalone system to cover certain process controls. The data transmitted from the DCU to the centralized operator console is not a duplication; it's process data through DCU.

Buchner: I'm only saying that you have several DCU's attached to your data highway, so that, if you have several direct control functions, several loops that you are controlling that are spread out, distributed amongst the DCU's, information on the process dataway goes from the process through one DCU onto the data highway to another DCU.

Kuroda: That is true. For example, in the case of cascade control, the master controller may be in the first DCU and the slave in the second. In that case, your data would go through this highway.

Yoshii: But I think that it is not a very good application design. You should pack all of the related loops in one DCU. So we do not recommend such a structure that requires direct transmission between two DCU's.

STRUCTURE OF AN IDEAL DISTRIBUTED COMPUTER CONTROL SYSTEM

Th. L. d'Epinay

Brown Boveri Research Center, CH-5405 Baden-Daettwil, Switzerland

Abstract. The problems related to distributed systems belong in general to the crucial ones in the design of computer systems. A general model which simplifies an efficient design of such systems has therefore to include appropriate means to deal with distributed systems.

Such a general model is presented here and it is shown, that a high level system, which provides virtually unlimited parallel processing capability independent from geographical distribution can in two steps be mapped to a low level system which only knows sequential processors.

Keywords. Distributed systems; communication; standardization; synchronization; portable software; levels of abstract machines; refinement and transposition; context switching; expandability; graceful degradation.

INTRODUCTION

Increasing efforts are made to reduce the costs of a computer control system over its whole life cycle. Appropriate methods are used like high level, block oriented languages, modular, portable and re-usable software, etc. on one hand and standardization of hardware components or interfaces between them on the other hand.

Doubtlessly such efforts have brought considerable benefit, but in the field of computer control systems a real breakthrough, comparable to the development in conventional computer systems, could never be achieved.

It is therefore worthwhile to analize the reasons of this unsatisfying development. In a simplified way the result of such an analysis shows two interrelated main points:

(1) The degree of freedom in the design of a computer control system is by an order of magnitude higher than in conventional systems,

(2) there is generally a high correlation between the methods chosen to solve a problem and the target computer system, on which the problem is solved.

Unfortunately the negative aspects of these two points are considerably increased, if the target system is *distributed*.

The consequence of point (1) is that each problem allows individual solutions, and human nature will insure that full use of that possible individualism is made. No standards (alone) will ever be able to reduce that sort of problem: the only effect will be that individual solutions will be expressed in common languages and use standard interfaces.

Point (2) shows that a "portable software" is a contradiction in itself as long as the method of the solution of a problem is dependent of the hardware of the target system, even if this solution is coded in a completely standardized language.

The conclusions which must be drawn from that short analysis are clear: A method to handle computer control systems - especially if they are distributed - must reduce the degree of freedom in the design of the systems and create independent levels between problem specification on one hand and the target system on the other hand. Obviously the reduction of the degree of freedom and the creation of independent levels is a very delicate problem, because the efficiency of an individual solution should not be decreased.

The research project presented here takes into account the above mentioned conclusions: A global concept is aimed at where possible specifications and standards are the consequence of the concept and not vice-versa. It is important that the problem of *distributed systems* is treated as a key problem within this concept.

GLOBAL CONCEPT

Description of levels

The description of the global concept is based on a very simple model (Fig. 1):

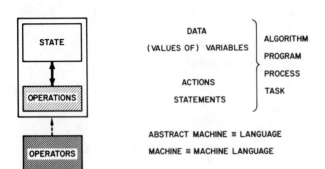

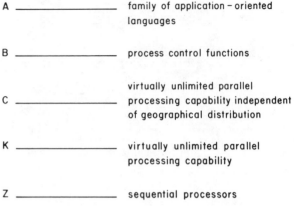

Fig. 2. Main levels and their characteristics

Fig. 1. Model to describe a system

Every system, especially a computer control system can be described in terms of a *state* which is influenced or changed by *operations*. In computer science a state is represented by "data" or "values of variables" and the operations by "statements" or "actions". Both together are called "algorithm"; expressions like "program", "process", or "task" may also be used. The set of possible data structures and the instruction repertoire chosen to describe the system represent one "operator", i.e. something which is able to perform the operations on the state. Again in computer sciences an operation is called machine or abstract machine which is identical to its corresponding machine language or language. This model may be applied to the whole range of possible representations of a control system, i.e. on one hand to the (implementation independent) specification of the problem and on the other hand to the actual (problem independent) target system. As the specification and the final implementation with a certain target system obviously should correspond to each other, these two representations are just two different images on different *levels* of the same problem.

The global concept distinguishes a number of such levels (Fig. 2), the solution of a problem will then be found, if the laws of projection of each level to its neighbours is defined. The connection between two levels is called *layer*.

Before we present further the nature and relationship between the different levels of the model, two basically different possibilities of relationship between two levels must be discussed (Fig. 3).

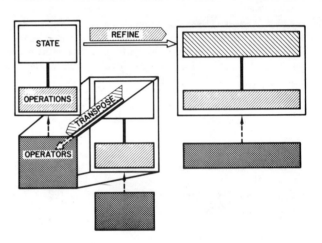

Fig. 3. Two methods of transformation: refinement transforms the algorithm, transposition transforms the operator

The method of refinement

This method is well known and described in /1/. It consists of a combination of a top down approach, subdividing more general algorithms of a higher level in partial algorithms which fulfil the solution of the higher level algorithm and represent elements of an available lower level target operator; and a bottom up approach, where elements of an available operator are combined in order to fulfil a required algorithm on a higher level.

The method of refinement is already widely accepted and used in many different forms, but in most cases it is used in an environment where only the target level (often a programming language) is specified, whereas the higher levels are left to the individual systems designers intuition.

The method of transposition

It is the aim of a transposition to map the *operator* (machine) of a higher level to a lower level (the refinement method maps the *algorithms*).

The transposition is a much more powerful tool and it allows the connection of levels with basic structural differences which usually cannot be connected by the refinement method. It also implies that the algorithm of the lower level does not directly correspond to the algorithm of the higher level, which among others influences the way of error and exception reporting from lower to higher levels. The main levels of this model are all mapped to each other using a transposition; the layers themselves are substructured using the refinement method.

THE MAIN LEVELS

The main levels are characterized by increasing capabilities with respect to parallelism, distribution and data-access.

Z-level

The Z-level is the lowest level considered in this model. Its definition is intentionally kept in general terms to allow its mapping to a wide range of existing and future hardware.

Three sublevels of increasing functional capabilities (Z1, Z2 and Z3) are used for a stepwise more precise functional description of the main layer on top of the Z-level. They also provide the possibility to map a more complex hardware to a higher sublevel.

The Z-level operator is defined to consist of any number of simple sequential processors. The only synchronization requirement is a consistent access to single-bit data shared between these processors. This facility is provided by the arbitration logic below the Z-level. Consequently, the Z-level program consists of a corresponding number of simple, sequential subprograms which are executed in parallel.

Although the Z-level is very simple, it can be described using a highlevel, block structured, sequential language. Thus the programs built on top of the Z-level with the sublevels Z1, Z2 and Z3, which together realize the next higher K-level, can be defined as programs in such a language. At present the whole layer is specified in the form of PASCAL-procedures. This allows extremely fast implementations of the K-level, even if it finally has to be done in µ-code or assembler. Test-implementations have shown that multi-processor K-machines have been implemented within 50-150 man-hours.

K-level

The K-level provides virtually unlimited parallel processing capability within processing nodes consisting of processor-pools which are able to run any number of co-operating *processes*. The instruction-repertoire of these processor-pools is not restricted, but all instructions by which processes influence each other are identical. They form a common subset of instructions for synchronization, data-exchange and process-management which is called *kernel*. It is specified and its realization based on the Z-machine is defined in detail. The selection of the kernel instructions and their realization needed a lot of special studies, modelling, simulation and testing. Although there exist a lot of proposals for synchronization and process management functions, the corresponding realizations are not suited to run on a multiprocessor hardware. The same is true for the structure of existing real-time operating systems, which do not consider the simultaneous execution of different operating system functions and have a totally unstructured internal data base.

In addition, existing proposals tend to be biased, whereas the kernel has to meet many different requirements, some of which are discussed here.

Reduction of the degree of freedom. This requirement is fulfilled by minimizing the number of kernel instructions. It has been proved that only two synchronization instructions are necessary to realize all presently known synchronization concepts. These two functions fulfil also all subsequent requirements.

Optimal hierarchy of synchronization functions (Fig. 4) The most primitive synchronization function is realized by the arbitration logic below the Z-level. The Z-level itself uses a pair of functions (LOCK (v) and UNLOCK (v)) in order to achieve mutual exclusive access to data. A variable (v) is associated with the different data structures, so that only functions accessing the same data are exclusive.

As parallel hardware is used, much care must be given to structure data and functions in a way to minimize the probability of access-conflicts. It is a fact that access-conflicts below the K-level lead to busy-wait states of processors and thus reduce the efficiency of the system. (If during the execution of a function which allocates a processor to a new process an access conflict occurs it is obvious that the processor cannot be re-allocated to avoid busy waiting). As the K-level is the lowest level which avoids busy-wait states of processors, this level may not contain any unnecessary complexity.

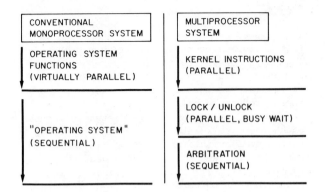

Fig. 4. Hierarchy of synchronization functions

In conventional systems virtual parallelism is achieved with a sequential operating system on a sequential hardware. In a multiprocessor system an additional level is introduced, where synchronization conflicts result in busy-wait states of processors.

Proof of absence of time-dependent errors and deadlocks. The combination of synchronization functions with functions of process management in a system which requires a maximum of parallelism causes a lot of new problems. As an example the following can happen simultaneously: (1) a blocked process A receives information from a process B and is switched to a runnable state, (2) process A is suspended by a third process C, and (3) the processor responsible for the execution of A recognizes the need to switch to a higher priority process D.

The absence of time dependent errors and deadlocks of such a system cannot be proved by testing or evaluation of all possible states. It is therefore necessary that all K-level instructions can be realized observing a set of simple rules and a strict modularity.

Efficient context switching. Several kernel instructions include context switching, i.e. allocation and deallocation of processors to processes. In order to make maximum use of potential parallelism and primarily to achieve maximum modularity extensive use of parallel processes is made.

Inefficient context-switching mechanisms are the reason for many cumbersome or non-modular functions presently in use; some of them are so generally used that one is tended to accept them as "basic" requirements of real-time and computer control systems. An example of such a detour-solution created to avoid "unnecessary" context-switching are the multiple wait functions (typically used in an input/output process which controls several devices). They can always be replaced by parallel process (typically one process per i/o device) executing simple wait functions. The interrupt mechanism is an other feature which one is inclined to accept as a "must" in a system with fast responses, but also this mechanism with its known sources of errors can completely be avoided.

The scope of this paper does not allow to prove in detail how a fast context-switching mechanism allows the replacement of cumbersome functions and thus leads to simpler and better structured systems. But it is evident that such a mechanism is a basic requirement for the K-machine.

High level languages. The reasons to support high level, block structured languages do not have to be explained here. With respect to the K-level the following points have to be considered:

- The concept of "context" with a slightly restricted meaning compared to the context discussed in the previous paragraph is also used in (sequential) block structured languages. Therefore these two notions of "context" must be treated as one problem and then *one* efficient mechanism will be sufficient.

- The rules to access variables defined in block-structured languages allow only a limited part of the memory to be accessed. If the addressing mechanism of the K-machine is defined to directly support these access rules, it yields different benefits: access-rights are enforced at run-time, a practically unlimited memory can be accessed with relatively few address-bits, context-switching is simplified and access to data is generally faster.

Dynamic expandability and graceful degradation. The K-level is a well defined machine built on top of multiple lower level machines. One of the reasons why this structure has been chosen lies in the fact that this consequent separation of levels allows the construction of the K-level in a way, that additional Z-processors increase the computation power of the K-level without the need to change programs. If special precautions are taken such extensions (or reductions) can be performed on-line and, with additional precautions, even uncontrolled reductions (i.e. failures) will only lead to a reduced speed of the K-level (graceful degradation). Our precondition to optimally support this behaviour requires that the local context of a process should not reside in processor-specific storage (where it is lost on a processor-failure). Ideally not even the program-counter should be located within a processor, but only an identifier referring to the actual block (of a process). This requirement fully coincides with the need for fast context-switching; a processor can be looked at as a "motor which can be "plugged in" to a "device" (process).

Of course, not only failures of processors, but also of other components of the system (busses, memories) must be considered. These problems can mainly be solved with known methods, like error-correcting codes, and are not discussed here.

Kernel instructions. The consideration of the last paragraphs lead to a small, simple and powerful kernel instruction repertoire, which is listed below:

Synchronization

INC (mailbox, information)
send one information to a mailbox. If at least one process is waiting for information of that mailbox, remove one process from the waiting list, pass the information to it and switch it to the active state *runnable*; otherwise insert the information to the information-list.

DEC (mailbox, information)
receive one information from a mailbox. If at least one information is available, remove one information from the information-list and continue; otherwise insert the requesting process to the waiting list and switch it to the active state *blocked*.

Process management

START (process)
A process is started synchronously; comparable to a block entry. If the process is not suspended, it is switched to the active state runnable.

RETIRE
The executing process has reached an end and is switched to an *inactive* state, comparable to a block exit.

SUSPEND (process, mailbox)
The system is told to switch the process (asynchronously) to an inactive state, if that is achieved, an appropriate message is put to the specified mailbox.

CONTINUE (process)
A suspended process is allowed to continue; if it is also started, it is put to the appropriate active state.

The detailed realization of the K-level on top of the Z-level is shown in /2/.

Functions of the K-level outside the kernel. For functions which do not belong to the kernel a certain classification with guidelines for each class is also envisaged. The main classes distinguished so far are execution control, algorithmical, input/output and data base management. They are not further discussed here.

C-level

The C-level consists roughly of functions with the same semantics as the K-level, but the objects of all functions may as well reside in the local as in remote nodes. As the K-level has been defined to include a minimal and optimal set of functions for synchronization, process-management and data-exchange there is no need to invent new semantics for the C-level.

The independence from the geographical distribution of processes and data objects is achieved with a number of very simple *communication-processes* in each node. These processes consist of execution control, kernel and i/o instructions. They consist typically of some 3-10 instructions. Thus, the C-level is completely defined and its implementation on top of any number of K-machines is documented.

The C-level hides completely all device-oriented i/o- and communication instructions, so that this whole class of functions is not visible anymore.

The most outstanding quality of the C-level is its simplicity which is of course only realizable based on the K-level. Considering that presently the switch from a centralized to a decentralized system requires in most cases a total redesign of the system, a solution where the logic of this procedure is described in a short and standardized form represents a major step towards problem- and solution-independent system design.

A system in which objects are accessed independently of their geographical position requires functions for routing, generation of protocols, etc. on lower levels. The model described here allows the inclusion of all existing communication functions at appropriate lower levels. It has become evident that *associative features* in the underlaying communication system greatly enhance the system performance and are specially valuable for on-line system reconfiguration.

Higher levels

At present the model is extended to include appropriate levels and transformations up to an application-level (A-level), which corresponds to a family of application-oriented languages, including graphical and semi-graphical representations. The transformations from the A-level to the C-level will not be fully automatic, but CAD methods will be used.

CONCLUSION

A complete model is presented which allows to treat system specification, system design and realization on a target system as a number of transformations between well-defined levels. This model starts on a very primitive lowest level consisting of simple, sequential processors. The next main level introduces means for parallel processing with the necessary synchronization tools and the third main level includes all means to run processes, exchange data and for synchronization in a distributed system.

This means that a program written on this third level is not aware of geographical distribution and runs on a virtually unlimited, distributed parallel processor.

As this third level is not only an utopical "nice model", but a machine whose construction out of a number of sequential, conventional processors is defined, it contributes substantially to the primary goal for an efficient and re-usable design of distributed systems.

REFERENCES

/1/ Wirth, N (1975). Systematisches Programmieren. Stuttgart.

/2/ Lalive d'Epinay, Th. (1979). Up-to-date report, technical committee on real-time operating systems of the international PURDUE workshop on industrial computer systems. Ed. no. 7.

DISCUSSION

Miller: Could you clarify what mechanisms you have in mind for translating symbolically-defined objects in some language into their final form for determining hardware memory addresses?. Also, could you discuss the binding mechanisms for this address transformation and when they would occur?

Lalive: Our main aim is to develop a concept and the concept should support different ways of realizing this mapping. It means there should be no difference in the concept whether it is mapped, for instance, at compile time or execution time. If there were a difference, it would not be a general concept.

In the model we have not yet said how addressing is done or if some particular mechanism should be used; we just say if there is any way to expand an address or to make the association between a logical name and the physical name, for example, then our model will tell you where to put it in.

Anders: On the Z level units, I was trying to picture hardware, even though we are talking concepts. I can see that you might have a Z level unit to do a Fast Fourier Transform, but later you said]00-200 program steps for a Z level unit. That would really vary depending upon the type of problem, but it still fits into the concept. What size is the Z level unit, or how do you talk about that kind of thing? This is a universal plug-in card, in my hardware way of thinking.

Lalive: Within our layered model you can ideally realize the K-level (as the K-machine) by sequential microprograms on microprogrammable Z-processors. Such a microprogram will realize all the instructions of the K-machine including, for example, instructions for Fast Fourier Transforms. Of all possible instructions, we have only specified the kernel part in detail, consisting of instructions for synchronization, process management (e.g. create, activate, deactivate, start process), etc.

We know that it is not always possible to realize the K-machine with microprograms, especially with existing hardware. Therefore our model includes guidelines for a microprogram-type implementation and for a procedure-type implementation where the capabilities of the K-machine are added to the capabilities of an existing minicomputer by a number of procedures. As the layer between Z- and K-level is defined with 3 sublevels (Z], Z2 and Z3), it is easy to actually build the layer on top of the effective capabilities of a mini, which lay somewhere between Z and K.

The size of a microprogram-type layer would correspond to today's microprogram plus the code to realize the kernel instructions. A procedure-type kernel for a multiprocessor K-machine consists of some 300-]000 instructions of procedures per minicomputer.

Anders: I was thinking the Z levels were individual functional units; rather they are functions. Do I have a computer at K level?

Lalive: You could think of it as the following: You have a number of microprogrammable microprocessors, with their own memory and with their own program console; as you do microprogramming on each, you have a sequential microprogram. They run asynchronously from each other and they realize the computer on the K-level which is similar to today's mini, but it is more powerful, has more instructions, and includes automatically all the power of these parallel microprocessors.

Buchner: At the heart of your presentation is the question of reducing development costs and you are led to developing a methodology for realizing systems. My concern is not with realizability, but more or less with practicality, in terms of performance and cost, as well. Of course, an extreme example would be that of a Turing machine to emulate any computer, but no one does this because of cost and performance. What you propose in terms of performance is replicating units on a particular level, either in terms of throughput or in terms of reliability. But, in fact, these considerations, if you look beyond development costs to global costs or global efficiency, might not be optimal solutions in terms of minimizing costs or obtaining maximum throughput. I wonder if you would comment on that point?

Lalive: You could map every existing model to our structure, and then develop your problem according to this model where you have the rules for how to do it. As a last step, you map it back to an existing structure. Until now we have found no single rule in our method which in a specific case would be in your way if you wanted to achieve the most efficient solution. Efficiency will, for instance, depend on your cost of cable, of what is available in hardware, and what is the requirement. All of this will have to go into a more or less semi-automatic compilation down from the specification of the application to the target system. But whatever your goals will be, you will have a grid in which you can think. Independent, for instance, from the communications system, you can develop your application and subdivide and translate it into subfunctions, leaving out all problems of optimizing your communication system. And, if you find out at some later date that you have to move some part from one location to another because you want to reduce the traffic on the communications system, you are able to move the process, in the informatic sense, from one part of the distributed system to another without having to change anything in your programs down to the C level. Below the C level you also might not have to change your levels, but that depends on how you construct your distributed system.

Buchner: My question concerns the interdependence between two levels in terms of optimality, because that relates to cost directly. As an example, you might be in the process of designing a controller. Let's say you start from an application level, and working on down to a lower level, you get to the point where you are designing some particular control and you decide on a suitable algorithm. Then you go down a few more levels and you are looking at constraints affecting your application algorithm, for example, a sampling period. At that point, you realize that in order to be optimal you might have to go back and change your control algorithm to be better suited to the knowledge that you now have on the particular limits on your sampling intervals; so there is an interdependency between the two levels in terms of the total overall performance. That's really what my concern is.

Lalive: That's very, very possible and you could think of that process as a sort of backtracking; you try to find the solution and you may have to go back if it doesn't fit, but all this is supported by having these levels. We could think of a computer aided system where, when you have made your transformation from one level to the other, the system will tell you if it's complete, free of contradictions, and all that. I don't think that such a system will avoid the necessity to have good system designers, but it will help them and, if there are only changes in one or two levels, the rest can stay as it is.

Steusloff: I think I found a lot of our ideas in your general model and I think it is very worthwhile to discuss it. I would just give a comment and give an example from our system of how we solved this problem of efficiency versus level. As I told you, we use the programming language PEARL which is a high level language and which has some efficiency problems, just like a lot of high level languages do. One of the problems is that a block-oriented language needs a lot of data access transfer when changing the block level. We have a processor which is microprogrammable, and we started with a standard instruction set. When we found out that the speed of changing the block level was too slow, we added into the microprogram some functions which support that and the problem was solved. That's essentially what Lalive says, just pick up the problems and put it into another level, but the overall structure stays the same.

Lalive: I may show you this picture (Fig. D]) of a very rough model of how these refinement steps look in PASCAL. You see it is the typical structure of a backtracking program. For those who like it in flow charts, I have it also in this representation (Fig. D2). You can see that you can also have a sort of learning system because things which have already been done in earlier developments can be stored in a library to cut off branches that you have already developed earlier.

Johnson: I think it is also interesting to view this problem as one of performing a computation, subject to the availability of certain resources. From your discussion, it seems to me that, in order to decide at what level a computation must be performed and what resources are required, you need to specify some information about the timing that's allowed for it in a real time control application and about the reliability. Have you made any progress in specifying those factors in a way that can be translated into how to use the available resources?

Lalive: I have said to you that we are at this so-called "A" level (application level) where we are further away from reaching our goals than in the other levels. But, one thing which will be sure is that already the A level will have to include figures about response times, data rate, and required reliability, so that the specifications will be complete. Then with each step of mapping, you can always check if all the specifications are met. The problem is that today specifications leave a lot of things open and you add these items when you do your refining. You never read a specification, or at least normally not, where it is written that nobody may be hurt by the system. You just automatically assume that and the same is true with other things. So that's a problem: How can this level be complete? This must include, of course, exactly the points you mentioned.

Discussion

```
PROGRAM refine(input,output);
(* this not fully developed PASCAL-like program shows
   the basic steps of refining a problem *)

TYPE moduletype = .... ;
VAR module: moduletype;
    modulelist: "list of modules";
    success: boolean;

  PROCEDURE subdivide (module: moduletype; VAR success: boolean);
  VAR submodule: moduletype;
      strategies: "set of known strategies";

    PROCEDURE selectsubmodule(s: strategy);
    BEGIN
      IF "selection possible" THEN
        BEGIN
          submodule:= "part"(module,modulelist,s);
          module:= module "REDUCED BY" submodule;
          success:= true
        END
      ELSE success:=false
    END;

  BEGIN strategies:= "set of known strategies";
    WHILE NOT (module "EMPTY" OR (strategies =[])) DO
      BEGIN
        REPEAT s:= "next strategy from" strategies;
          strategies:= strategies-s;
          selectsubmodule(s);
          IF success THEN
            IF submodule "sufficiently refined"
              THEN print (submodule)
              ELSE subdivide (submodule,success)
        UNTIL success OR (strategies=[])
      END
  END;

BEGIN "initialize modulelist";
  READ (module);
  subdivide (module,success);
  IF NOT success THEN write('No solution found.')
END.
```

Fig. D1:   Refinement Steps in Pascal Program Fragment

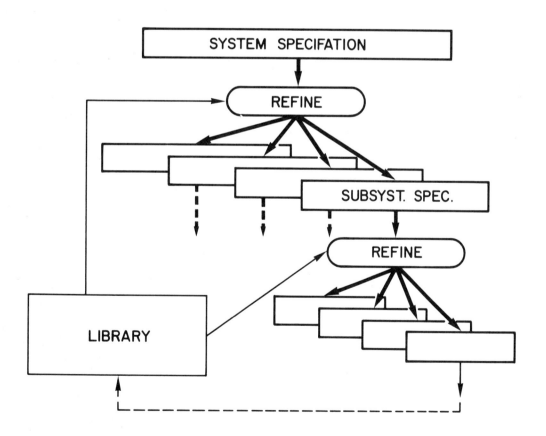

Fig. D2:   Flow Chart of Refinement Process

DISTRIBUTED DATA PROCESSING SYSTEMS — DISTRIBUTED COMPUTER CONTROL SYSTEMS: SIMILARITIES AND DIFFERENCES. A PANEL DISCUSSION

T. J. Harrison

International Business Machines Corporation, Armonk, New York, U.S.A.

Abstract. This paper introduces the five following papers which are the introductory statements made by the participants in a three-hour panel discussion. In this paper, the ideas leading to the formation of the panel discussion are presented and a hypothesis is stated to serve as the central theme of the discussion. In order to clarify semantic differences, a working definition of "distributed system" is proposed. In addition, three questions relating to specific areas of control mode, intercommunication, and system integrity are posed as a starting point for the discussion. The panel discussion was recorded and transcribed, and the edited questions, answers, and comments are included in these proceedings following the introductory papers.

Keywords. Computer architecture; computer control; computer organization; digital computer applications; digital computers; digital control; system integrity.

INTRODUCTION

A panel discussion is included as a part of this Workshop on Distributed Computer Control Systems. The panel consists of five individuals whose current employment represents a research institute, a university, an industrial computer vendor, a data processing computer vendor, and an industrial computer user in the oil industry. The individuals were selected to provide a balance between expertise in traditional data processing and expertise in the use of control computers in industrial applications.

Each of the participants has been asked to provide a ten-minute introductory presentation. The five following short papers published in these Proceedings are the written version of their introductory remarks. The papers were prepared independently, and none of the participants have seen the papers of the other participants prior to arriving at the Workshop. All panelists, however, were provided copies of this paper to assist them in preparing for the panel discussion.

In reading these proceedings, the independent preparation of the five short papers should be taken into account. Furthermore, the edited transcription of the discussion which follows the papers represents an important clarification and expansion of the ideas of the participants and other members of the Workshop. Thus, this set of six papers and the transcription should be considered as a unit.

The title selected for the panel discussion brings to mind an immense spectrum of possibilities for detailed consideration. The concepts of distributed processing are in their infancy, and precise definitions which would alleviate semantic difficulties do not exist. Our problem in dealing with this subject can be compared to capturing a cloud and transporting it for 1000 kilometers: It is difficult to tell where the cloud begins and ends, it is difficult to hang on to the cloud, and transporting it is drastically affected by outside influences such as atmospheric pressure and the turbulance of passing jet aircraft. But, let us grab our capture nets and see if we can solve our illusive problem.

AN INITIAL HYPOTHESIS

Our approach in this panel discussion is to state a hypothesis and then to explore its validity. The hypothesis is derived from this writer's personal experience in the computer business. Early in my career, my interests were almost entirely focused on what were called process control computers, computers which were intended to be interconnected with an industrial process for the purpose of controlling that process. With the advent of the mass-produced minicomputer (and now the microcomputer), however, it has become increasingly difficult for a computer manufacturer to concentrate its efforts in an application area as narrow as process control. Minicomputers and microcomputers today are used in almost all types of computer applications. As my interests have expanded from the process control area to a broader spectrum of applications, I have had the opportunity to deal with applications from direct digital control to payroll processing.

In reviewing my experiences, it is my perception that the users and, to a lesser extent, the manufacturers tend to cluster into vaguely defined "masses," not unlike cumulus clouds, which are semi-independent and only rarely seem to communicate effectively. Thus, there are the traditional data processing "buffs" who know or care little about industrial processes; similarly, the industrial control buffs are often unaware of the data base buffs who worry about content addressing, search trees, and the like. One can go on to define numerous computer specialties such as the computer game buffs, the scientific problem solvers, the artificial intelligence gang, and many more. Each specialty group feels that its problems are unique, and they seem rarely to look for solutions in the next cloud. The result, in my opinion, is that there is a significant amount of reinvention occurring and that the optimum solutions of one specialty have not been made available to the members of the other specialties.

The Hypothesis

Perhaps my view is incorrect or narrow. As a means of testing this, I propose consideration of the following hypothesis:

> The functional, physical, and performance requirements for distributed data processing (DDP) and distributed computer control (DCC) systems are substantially the same; the differences that exist primarily are matters of degree in the multidimensional parametric space which describes general-purpose computing systems.

Proving the Hypothesis

Proving any hypothesis is difficult. It requires that all possible cases be considered and that the hypothesis must be shown to be correct in every case. Many times, however, disproving the hypothesis is easy; a single example showing that the hypothesis is untrue constitutes a necessary and sufficient condition for rejection. I have suggested to the panel participants, therefore, that we concentrate on defining the similarities and differences between DDP and DCC systems. The similarities tend to support the hypothesis and, in a large number of cases, I expect that we can all agree with little discussion. The differences tend to disprove the hypothesis and are likely to invoke lengthy discussion. In both cases, I expect that a great deal of our consideration will be directed to "how similar" or "how different" are the requirements. Because of the spectrum of data processing and control applications, it will be difficult to quantitatively express the degree of similarity and difference. I would propose, therefore, that we think in terms of a continuum ranging from "identical" to "unique" with a numeric scale of 1 to 10.

Unique   Identical
———————————————————————————————
 1   2   3   4   5   6   7   8   9   10

It probably will be necessary to assign a range of values to a requirement that depends on the particular application. Thus, for example, it might be concluded that an interrupt response time of 10 microseconds is unique (=1) for counting nuclear particles and ranges to a value of 6 for "classical" process control when compared to interactive data entry.

The Results

The complexity of modern computer systems, the spectrum of applications, and the multidimensional space of descriptive computer parameters, coupled with the three-hour time limit on the discussion, means that the hypothesis will be neither proved nor disproved today. Knowing that success is an unlikely result of today's discussion, what value can be expected from today's effort?

I believe that the most important result can be an increased understanding by all of us of the tremendous intellectual resources which exist in the various subdisciplines of the computer and control sciences. Hopefully, this understanding will lead us to seek answers to our seemingly unique requirements from experts in other subdisciplines. Successfully utilizing these many resources can only result in better solutions in less time.

A FRAMEWORK FOR DISCUSSION

In reading through these proceedings and other recent computer literature, it is obvious that distributed systems are topics of considerable interest. It is equally obvious, however, that the words "distributed system" mean different things to different people. The lack of recognized definitions of DCC and DDP systems and their components has been, and will continue to be, a source of confusion and difficulty in communicating ideas about the systems and concepts involved. This lack is a result of the relative infancy of the concept. As part of the framework for our discussion, therefore, I believe that we should start with a working definition of a distributed system. From this starting point, the speakers and Workshop members can derive their definitions, as necessary.

Working Definition

The applicable definitions in Webster's Dictionary (1963) define the words system and distribute as:

System: 1) a regularly inter-
             acting or independent
             group of items form-
             ing a unified whole;

           2) an organized set of
             doctrines, ideas, or
             principles usually
             intended to explain
             the arrangement of a
             systematic whole.

Distribute: 1) to divide among
             several or many;

           2) to spread out so as
             to cover something.

Unfortunately, concatenating or otherwise combining these two definitions does not seem to clearly describe what we mean by a distributed computer system as compared to a centralized system. For example, a conventional uniprocessor-based computer system seems, simplistically, to satisfy the definition: such a system has functions assigned to various parts of the hardware such as the peripheral units, the central processing unit, and the channels. These units regularly interact to form a unified computing system.

A second problem with the definition is that it does not explicitly indicate what is divided among several or many: Is it the physical units which are geographically dispersed, the logical processing capability, the data, or the control? Or is it all of these? Clearly, we need to decide what it is that we are distributing in a distributed computer system.

Several years ago, Enslow (1978) proposed a working definition that had five components, all of which he considers necessary to uniquely define a distributed data processing system. The five components are:

"A _multiplicity_ of general-purpose resource components, including both physical and logical resources, that can be assigned to specific tasks on a dynamic basis. Homogeneity of physical resources is not essential.

"A _physical distribution_ of these physical and logical components of the system interacting through a communications network. (A network uses a two-party cooperative protocol to control the transfer of information.)

"A _high-level operating system_ that unifies and integrates the control of the distributed components. Individual processors each have their own local operating system, and these may be unique.

"_System transparency_, permitting services to be requested by name only. The server does not have to be identified.

"_Cooperative autonomy_, characterizing the operation and interaction of both physical and logical resources."

This seems to be a better definition in that it tells us a little more about what is being distributed and how these distributed elements act and are controlled. In his article, Enslow elaborates on each of these five components in more detail. In a companion article, Eckhouse and Stankovic (1978) report on two workshops which considered some of the issues related to DDP systems. The reader is referred to these articles for more information.

In summary, therefore, I propose that we accept Enslow's working definition as the base point for our discussion. I encourage the panel participants and the members of the Workshop to examine this definition and comment on its applicability to both DDP and DCC systems.

The Multidimensional Parametric Space

The hypothesis refers to a _multidimensional parametric space_ which describes general-purpose computing systems. Actually, there are many such spaces that can be constructed. For example, one might construct such a space for the general concept of _computer performance_ in which overall system performance is represented by a vector whose components represent all those system parameters which affect performance. Examples of such parameters are the speed and size of main memory, the functional capability of I/O devices and channel attachment hardware, the overhead of the operating system, the efficiency of the language translator programs, and so on. It is easy to see that this multidimension n-space is very large.

Similarly, large n-spaces exist for each aspect of DCC and DDP systems that we choose to consider. Furthermore, n-spaces can be constructed at various levels of detail. For example, processor instruction execution rate is a parameter of overall computer performance. But, processor execution rate can be described as an n-space having components such as the gate delay of the logic used in the hardware implementation, the number of registers in the architecture, the number and transfer rates of internal busses, etc.

Thus, in order to be able to comprehend the similarity and differences between DCC and DDP systems, we need to choose the appropriate n-spaces at the appropriate level of detail. For our purposes in this panel discussion, I believe that we must initially consider spaces of limited dimensions and with relatively course parameters.

Enslow (1978) characterizes DDP systems in a three-dimensional space that both illustrates the concept of a multidimensional parametric space and can serve as a starting point for our discussion. Enslow's figure, reproduced here as Fig. 1, characterizes a distributed data processing system in terms of the decentralization of hardware, control, and data base. He suggests that DDP systems occupy only a limited portion of this space, as illustrated in the figure. I suggest that this figure be examined to determine if we agree that these are the right parameters to examine and if DDP and DCC systems occupy the same subspaces.

SOME QUESTIONS TO CONSIDER

Given that the working definition is satisfactory for our purposes and that we must necessarily focus our attention on a limited set of parameters, I propose that the panel participants and the members of the Workshop consider the following three questions:

Peer Versus Hierarchical Control

Many of the current articles in DDP stress that peer control is a necessary requirement. In this concept, independent resources in the system are relatively autonomous and deal with other resources on an equal or peer basis. In human organizations, this might be exemplified by the manner in which unpaid volunteers in a civic organization interact with each other. On the other hand, some researchers in the area of DCC systems (e.g., Williams, 1979) feel strongly that a hierarchical delegation of authority is a necessity in computer control applications. In essence, they argue that there must be a system manager who is aware of, responsible for, and in control of, the entire system. You will note that Enslow's diagram (Fig. 1) seemingly excludes hierarchical control from the DDP space.

Communication Between Processes

In this context, the word "process" refers to a computing process, not the industrial process to which a computer control system is connected. The primary question is how to pass data and requests between independent cooperating computing processes in a distributed system. Two possibilities, perhaps at the end of a spectrum, are a message system and the use of a common or global shared access communication space. In the message system, all communication takes place by means of discrete messages between processes, even if both processes reside in the same physical processor. Conceptually, the method is analogous to a written mail postal service. The initiating process packages a request or set of data (puts it in an envelope) and places it in a known location (a mailbox).

The receiving process (which may or may not have been notified of the waiting message) is responsible for retrieving the message from the location. This requires cooperation on the part of both processes to ensure that the message is delivered on a timely basis.

The shared communication space, on the other hand, is analogous to a public bulletin board. Data and requests are publicly posted in a way that they are available to all processes. This has the virtue of simplicity but may suffer if the public notice is erroneous or altered in such a way as to propagate an error through the system. It also has the advantage of speed since only the location of the data in the common communication space need be transmitted to receiving processes.

The concensus of opinion reported by Eckhouse and Stankovic (1978) for DDP systems is that protocols for communication, and efficiency versus protection trade-offs are a more important distinction between interconnection methods than are questions of bandwidth and the degree of coupling between processes. A question, then, is whether this same concensus applies to DCC systems.

System Integrity

Many proponents of DCC systems insist that system integrity is one of the most important requirements in any system design. They argue that the detection of errors, automatic recovery, or at least the isolation of errors so that they do not propagate through a system, is essential to avoid catastrophic consequences for personnel and equipment if an industrial process should go out of control. As a result, they often insist on relatively sophisticated error correction coding, redundant communication paths, and even redundant backup systems.

However, system integrity is also important to the DDP user. An error introduced into a data message in an electronic funds transfer system could cause considerable embarrassment to the owner of an account which was overdrawn (or joy to the recipient of an undeserved credit of a large sum of money!). Similarly, inadvertent erasure of records for a day's receipts could constitute a financial crisis for a company. Thus, although the DDP user might not cite loss of property or life as a consequence, the need for system integrity might be considered equally great.

SUMMARY

In this short introduction, I have attempted to set the scene for the following five papers and the panel discussion. I have suggested a working definition and several questions to guide our initial discussion.

I now invite the panel participants and the members of the Workshop to express their ideas on these points and on others of their choosing that will allow us to examine the similarities and differences between distributed data processing systems and distributed computer control systems.

REFERENCES

Eckhouse, R. H. Jr., and J. A. Stankovic (1978). Issues in distributed processing -- An overview of two workshops. Computer (U.S.A.), Jan. 1978, 22-26.

Enslow, P. H. Jr. (1978). What is a "distributed" data processing system? Computer (U.S.A.), Jan. 1978, 13-21.

Webster's Seventh New Collegiate Dictionary (1963). G. & C. Merriam Co., Springfield, MA.

Williams, T. J. (1979). "Micro-Modules: An overview of Industrial Requirements." Presented at IEEE Computer Society Workshop on Microcomputer Firmware/Software, Applied Physics Laboratory, Laurel, Maryland, March 7, 1979.

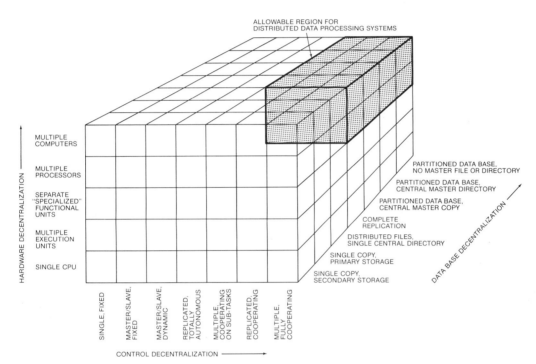

Fig. 1. Dimensions characterizing distribution

INTRODUCTORY REMARKS: DDP VS. DCC

T. L. Johnson

Laboratory for Information and Decision Systems, Massachusetts Institute of Technology, Rm. 35-210, Cambridge, Massachusetts 02139, U.S.A.

INTRODUCTION

Certain similarities between existing distributed computer control systems and distributed data processing systems are apparent, even though these two classes of large computing systems find their origins in different sorts of applications. The main question I should like to raise, is whether these apparent similarities are merely a consequence of common implementation constraints for both application areas (e.g., currently available hardware and software) or whether they arise from some deeper conceptual similarities between the problems being solved.

APPARENT SIMILARITIES OF EXISTING DISTRIBUTED COMPUTER CONTROL AND DISTRIBUTED DATA PROCESSING SYSTEMS:

Great similarities can be identified in the architecture of distributed control and data processing systems. Two common patterns are identified in Figure 1: (a) in a heirarchical architecture there is a large central processor which handles all communication and coordination between a set of local processors; (b) in a decentralized architecture, there is a communications network interconnecting a number of local processors, and the management and coordination functions are also decentralized.

A variety of configurations intermediate between these extremes can be found; perhaps (a) is relatively more prevalent in distributed control, while (b) is relatively more prevalent in distributed data processing. The sort of examples we have in mind for distributed computer control arise in metallurgical and chemical refining and manufacturing, while prototype examples of distributed data processing might include transaction processing systems such as banking, airline and hotel reservations, inventory management and shipping, automatic text retrieval for libraries or newspapers. The software features of these applications areas appear more diversified than the computational hardware, but certain common issues such as multiprocessing, virtual memory, and communications protocols should be noted.

CONTRASTS BETWEEN CONCEPTUAL DESIGN PROBLEMS IN DISTRIBUTED COMPUTER CONTROL AND DISTRIBUTED DATA PROCESSING SYSTEMS

In asking whether the apparent similarities between these two types of systems arise from underlying features of the applications they originate from, one common aspect of the applications is evident: At spatially distinct sites, there is a need for digital processing capacity, but the processing at the various sites must be somehow coordinated. Let us contrast the specific characteristics of the local processing and coordination requirements, however.

In a control problem the local process, or part of the plant at each site, is relatively well-defined by engineering considerations. In particular, the time-scale of the local process determines the required bandwidth of the controller, and the ways that the local process interacts with the rest of the plant (e.g., flows of material or goods) is well-defined. The type of information required from other parts of the plant and the frequency with which it must be updated, are also reasonably well-defined. If the interconnecting structure of the overall physical plant can be modified at all, the number of possible modifications is usually finite and they are not frequently made (relative to the time-scale of the local process). As a consequence of these factors, the communications requirements between parts of the plant are relatively well-defined, and the communication protocol may even be optimized for the particular application. Whether implementation is synchronous (e.g., sampled-data or asynchronous (e.g., event-driven), the real-time requirements of the problem must be satisfied. The reliability of the over-all distributed control system must be high, and the local processors must be extremely reliable, as there is a potential for instability in the event of malfunction.

A further consequence of the requirements of a distributed control problem is that the local CPU spends a relatively large proportion of its cycles in computation and input-output operations; often the local memory requirements are modest, and less time is spent in memory referencing. By contrast, in a distributed data processing application, the local CPU might spend a much greater proportion of its cycles in memory-referencing operations, and have correspondingly greater storage requirements at the local level.

In a distributed data processing system, the basic requirements of a typical application are often quite different from those of a distributed control system. The local processor in many cases might be an "intelligent" terminal interacting with a human operator; this form of feedback is evidently very different from that previously discussed. Typically the operator will draw on, or contribute to information in a local data base, and (perhaps less frequently) may exchange information with more remote parts of a system data base. If the operator is viewed in the same terms that we previously viewed the local process, as part of an overall "plant", his requirements are very different and everything is less well-defined: the time-scale of his interactions, the format and scope of his interactions with other parts of the system-even the sort of information he requests from his own data base. As a consequence, the stochastic characterization of the operator will dominate the structuring of the data base as well as the communications protocol of the network. Usually, these dictate greater flexibility and robustness than in a distributed control system. Even the structure of a distributed data processing system may change dynamically, on the same time scale that local transactions take place, so that versatile routing and coding strategies may become important. The reliability requirements in distributed data processing are also of a different nature than in distributed control. While the malfunction of a single node or link may not have catastrophic consequences, the reliability of the data storage and the integrity of the network as a whole are often extremely important.

CONCLUSIONS:

From the foregoing discussion, the similarities between distributed control and data processing systems appear relatively superficial, while the fundamental differences are very great; the different approaches taken in these different problems are well-justified. However, it is possible to formulate a distributed data processing problem as a control problem, and vice versa; this is useful to bear in mind, because most problems in fact fall on a continuum between the extremes that have been discussed. For instance, distributed control systems might benefit from more flexible communications protocols and more attention to data base structure, whereas distributed data processing systems might benefit by accounting for certain real-time processing requirements or by taking into account the feedback via dynamics of the socioeconomic and institutional transactions with which they deal. In the future, we would expect to see increased diversification in the means of implementing solutions of these large-scale computing problems.

Acknowledgement: Support for participation in this workshop has been provided by the National Science Foundation Engineering Division, Systems Theory Program, under Grant 77-28444.

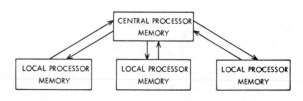

(a) Heirarchical Architecture

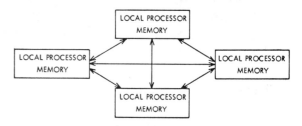

(b) Decentralized Architecture

Figure 1: Configurations Common to Distributed Control and Distributed Data Processing

TRANSACTION PROCESSING - AN EFFECTIVE CONCEPTUAL FRAMEWORK FOR COMPARING AND UNDERSTANDING DISTRIBUTED SYSTEMS

W. L. Miller

Fischer & Porter Company, Warminster, Pennsylvania, U.S.A.

Abstract. This paper describes the introductory statements of the author presented in a panel discussion that compares the essential characteristics and requirements of distributed data processing (DDP) and distributed computer control systems (DCCS). The similarity hypothesis and definition of a distributed system set forth in the introductory paper by T. J. Harrison are supported. The parametric measures of control mode, interprocess communication and system integrity are discussed. Finally, the opinion that transaction processing is an excellent frame of reference for comparing and understanding distributed systems is postulated.

Keywords. Computer architecture; computer control; computer organization; digital computer applications; digital computers; digital control; system integrity.

INTRODUCTION

Distributed Systems, both distributed data processing (DDP) and distributed computer control (DCC) systems are assembled from common components that form a similar kernel for both DDP and DCC systems. The kernel involves the technology of operating systems, compilers, data bases, computer hardware, and data communication networks. Distinct but superficial differences exist between DDP and DCCS in human and process interfaces. It is my opinion that the important differences, if any, exist in the kernel and are a result of differences in requirements.

The definition of a distributed system given by Enslow (1978) is wholeheartedly accepted and paraphrased as follows to further support the similarity hypothesis. Both DDP and DCCS must exhibit similar characteristics to qualify as a truly distributed system. The requirements are:

1) Many, locally autonomous operating systems, computers & processes, but with mutual cooperation on system tasks.

2) One integrated data base, physically distributed, but with data location transparent to the user. Items optionally accessible by symbolic name only. Minimal copies, except for back-up.

3) One integrated data communications network, optionally redundant, and transparently invoked as needed by the system to access remote data for the user.

4) On-line reconfigurable processing, network topology, and data base.

5) Constant user interface for a system configured as either distributed, multiple processors or a central, single processor.

As an example of the applicability of the above definition, a recently introduced DCCS (DCI-4000®) meets all of the above requirements in the 3 dimensional, parametric space proposed by Enslow. DCI-4000® has 1) multiple computers, all locally autonomous; 2) multiple control points mutually cooperating on the execution of partitioned system tasks and 3) a partitioned data base with minimal data copies but several central master directories.

SYSTEM DESIGN CONSIDERATIONS

To focus on the comparison of the DDP and DCCS kernels, several basic apriori design considerations will be stated. These are:

1) Decomposition of a system into areas or spatial regions of high data flow enables processing performance and operational security and efficiency.

2) Integration of related functions into autonomous units enables flexibility of system configuration and operational reliability.

Peer Versus Hierarchical Control

A study of the requirements of a distributed computer control system will reveal several factors that contribute to a justification for hierarchical structure. One is that DCCS applications have close corresponding relationships between response time and scope of operational responsibility, and that these are aligned with the flow of information upward from the process through the user corporate organization. Another is the practical efficiency of limiting the square-law potential communication paths between all low-level units. These and other factors cause an <u>external</u> DCCS structure (DCI-4000®) to be both a spatial and functional hierarchy as shown in Fig. 1 and 2. However, the modular autonomy in a distributed system that enables configuration flexibility, incremental growth, high availability, high reliability, expansion in capacity and function and other desirable features requires an <u>internal</u> system structure that is <u>based on</u> a peer-to-peer protocol between local operating systems and other critical components in the system kernel. Also, whereas the DCCS structure may have several hierarchical layers, peer control is the mechanism within an individual layer. Therefore, a DCCS uses a hybrid combination of both hierarchical and peer control to meet its requirements.

Communication Between Processes

As was stated in the introductory paper, Eckhouse and Stankovic (1978) relate a consensus that for DDP systems, protocols and efficiency considerations are a more significant distinguishing factor between interconnection methods than is bandwidth. Extending this concept, in my opinion, it's <u>the level of information</u> transferred or processed by any given number of messages that is important and determines the relative overall system response and throughput. The effective utilization of bandwidth is what's important, not the absolute raw capacity of a channel.

In order to compare the levels of information transferred in different systems in a common conceptual framework, it's helpful to define the concept of a process transaction. A process transaction is defined to be the gathering of atoms or parameters from accessed records in the distributed data base, the processing of the data and the updating of record atoms, again in the distributed data base.

One DCCS (DCI-4000®) has great similarity to DDP transaction processing. Both are real-time activities. In the DCCS (level 1), the response time is determined by the process, whereas in DDP and DCCS (level 2) it's typically determined by human interaction at a CRT terminal.

By using the concept of a transaction, the internal data flow and operation of a DCCS can be better understood. Whereas, at level 2 within DCCS, messages may be explicitly used for data transfer, at level 1 in a DCCS shared memory may be directly accessed with the explicit message mechanism only logically simulated. Therefore, a DCCS probably exhibits a hybrid combination of interprocess mechanisms, in this case explicit messages and shared memory to operate effectively.

System Integrity

The consideration of data communications integrity involves for example, the undetected error rate, redundant paths, ambiguity of message sequencing, buffer allocation lock-up, error recovery, infinite wait prevention, and priority response algorithms. These same issues must be considered in either a DCCS or DDPS.

The consideration of data base integrity involves, for example, the indivisibility of update operations and coordination of concurrent updates initiated by multiple, independent processes, the lack of a centralized locking mechanism, deadlock prevention, undetected configuration changes, and data base roll-back and failure recovery. These issues are complicated by the need to do system control without having global, deterministic state information.

But, they all have to be handled in both a DDPS and DCCS. Utilization of the transaction concept contributes to an effective analysis and determination of system integrity in a DCCS.

SUMMARY

There is a great amount of similarity between DDP and DCCS and the examination of the similarities and differences can be considerably enhanced by using the conceptual framework of transaction processing.

DCCS DECOMPOSITION

A REGIONAL HIERARCHY

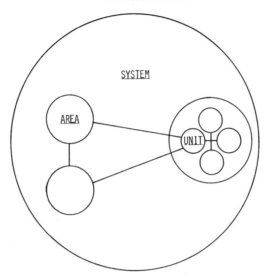

SYSTEM OF AREAS OF UNITS

FIG. 1

DCI-4000® - ONE AREA
FUNCTIONAL HIERARCHY

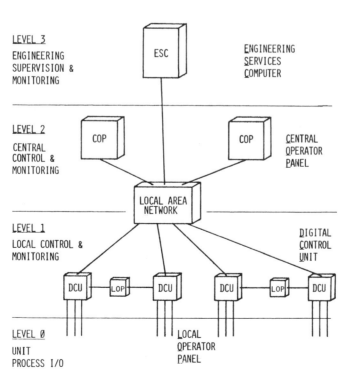

LEVEL 3 — ENGINEERING SUPERVISION & MONITORING — ESC — ENGINEERING SERVICES COMPUTER

LEVEL 2 — CENTRAL CONTROL & MONITORING — COP — CENTRAL OPERATOR PANEL

LEVEL 1 — LOCAL CONTROL & MONITORING — DCU — DIGITAL CONTROL UNIT

LEVEL 0 — UNIT PROCESS I/O — LOP — LOCAL OPERATOR PANEL

FIG. 2

A VIEW OF BUSINESS REQUIREMENTS ON DISTRIBUTED DATA PROCESSING SYSTEMS

R. P. Singer

*Systems Communications Division, IBM Corporation,
1000 Westchester Avenue, White Plains, New York 10604, U.S.A.*

Abstract. In introducing this author's view to the discussion of similarities and differences between distributed data processing systems and distributed computer control systems, this paper discusses some of the key requirements in the business environment highlighting the need for some degree of central control over the DDP environment.

Keywords. Central control; distributed systems; operating systems stability; operational control; software modification.

INTRODUCTION

Tom Harrison has assembled a distinguished panel today, and I am pleased to be a member of it. My background is primarily in computer systems analysis. For the past five years I have been working in the area of distributed data processing.

My experience in control systems dates from over a decade ago when I was involved in a number of projects in the Midwest ranging from the central control of a very large water filtration facility to the investigation of the requirements for, and proposal of a system to directly control a pilot coal gasification project. I worked on numerous systems that attached quite a range of laboratory instruments and one system that was my first venture into multiple, interconnected computer systems. This latter system had some characteristics similar to what I consider a distributed computer control system. Since it took place a dozen years ago I guess I ought to review some of the key characteristics for such an early attempt at a DCC system, an attempt that predates, to the best of my knowledge, the label of "distributed" itself.

The installation was in a research and academic setting that was performing dozens of experiments covering quite a range of physiology and medical areas of investigation. The funding for different experiments came from different sources, with substantial portions for many of these projects coming from different government and non-government grants. This sometimes contributed to the trend to a distributed implementation. While there was continuity for many of these projects, there were considerable changes, with some projects being discontinued, and new ones starting frequently.

Numerous computers of different sizes, of different architectures and from different vendors were interconnected to support these experiments. While the loadings on these systems affected the operations such that some experiments had to be scheduled in advance, and precluded certain others or combinations of experiments from concurrent operations, the general facilities supported multiple experiments concurrently. "Off loading", the term we used at that time, occurred for reasons of performance/response time, simplicity of operation, availability of time and/or equipment, economics as well as simply the preferences, knowledge and familiarity of the implementors.

Needless to say, much of the support for that system was what we would today consider primitive, but recall that the vast majority of the software was provided by these users, and not the vendors. One further caveat is warranted; that group of users was the brightest, most competent group of computer people I have ever encountered.

This brings me to a key difference between this installation and a normal commercial environment - of either DDP or DCC. The end users of that system were active users, utilizing the DCC system to develop their own autonomous applications utilizing distributed resources and functions. DDP's use in the commercial environment is primarily for developed, installed and operating applications to support non-professionals. Although some autonomous development may occur, it is generally quite limited. I distinguish, then, between the customer's requirements, that is the requirements dictated by the enterprise's operations, from the ultimate end user of the DDP system.

As a result of this view, one of the areas which we have found necessary to stress is the idea of centralized control over the distributed data processing environment. The customer must be able to effect some control over the disparate elements in his system.

CENTRALIZED CONTROL

This view differs from Philip Enslow's definition. In his definition Enslow appears to envisage a series of co-operative peers, each capable of carrying out a set of functions and able to route requests for additional function to some peers, capable and willing to perform the service. Although this view may be satisfactory for academic or research environments, I believe it overlooks some basic pragmatics of the commercial environment.

Businesses are organized in hierarchies of groups, departments, business units, divisions, etc. These organizational structures dictate a solution that at least has the potential to reflect the hierarchy. That implies that there may indeed be a logical master/slave relationship, at least at some level of the application. Further, different organizational entities have different areas of control. This again dictates that the function association cannot be performed by the elective capabilities of equal peers, but must be controlled and enforced by some higher level of organization, in one or more centralized processing facilities.

This overall system control can oversee and balance work loads, can ensure commonality of function within and among the organizational entities, and can control the spread and utilization of individual components. Given this view, there is a further necessary limitation that the centralized facility must be able to communicate with all elements in the distributed processing environment. Current pragmatics of technology suggest that this control and communication can be more efficiently realized with a limited set of similar operating systems in the distributed nodes.

I will now introduce some more specific functions that mandate the view of some level of centralized control in the commercial environment. Operating systems and some application support are provided by the vendors of distributed processing systems. It is unlikely that the function supplied by the vendor will precisely meet the requirements of every customer. Remember, I am talking of the customer, not the end user of the distributed processing system. Thus customers will make modifications to these vendor supplied functions. There is a need to ensure that the implementation and testing of these modifications is controlled and properly distributed. Modifications must be done in a uniform manner among all the appropriate participating nodes.

Data distribution and its control is another area where the customer requires coordinated management of the distributed data base. Individual data bases can be set up and maintained to reflect the needs of the autonomous nodes, however some requests are outside the domain of the local node and require the intervention of some higher intelligence to resolve. Then communication occurs between the local node and some central facility. Such communication should occur transparently to the end user.

Even in the area of general purpose application development, the customer needs a facility to control the orderly development of the applications in the distributed processing environment, as well as to monitor the installation of these applications in the participating nodes, thus ensuring consistency of function among the participating nodes. The presumption again is that in the commercial environment, the customer wants the benefits of implementing and distributing applicatons and does not necessarily want, nor have, the skills at each distributed site to permit remote customer personnel to implement their own applications.

Another area of prime importance is operational control. Clearly it is economically unfeasible for individual distributed processing nodes to support their own systems operator.

Equally clearly, it is necessary, at times, to call on the particular skills of a systems operator. The reconciliation between these two anomalies is to provide a centralized systems operator who can access and control the distributed node. This access would be to install corrections to software, to investigate error conditions, and to perform debugging analyses. It must, of course, be implemented in such a way that it occurs without disruption to the user at the distributed processing node.

Some level of centralized system control is also required in the areas of reliability and error recovery. It is unreasonable to expect the user of the distributed processing node to be concerned about the reliability or to determine the levels of problems occurring within the network.

In summary, distributed data processing in the commercial environment requires an overall systems control in the areas of:

- modifications and updates,
- resolving data accesses,
- operator and network control.

As such, DDP can also reflect the hierarchies of the particular business environment.

FUTURE PROBLEMS

As we move further into the field of distributed data processing, we realize that we can and must make some considerable strides in some new areas. We are now supporting the business organizations of our customers in an entirely new way, and they are relying on the continued provision of this support. In some cases there have been reorganizations to better utilize this support. Thus it is necessary that the operating systems in the distributed processing environment be very stable. Although modifications will occur, both for software corrections as well as to provide new functions, these modifications must be made available in a controlled manner such that the customer can upgrade only as his business requirements and economics justify. Such upgrades in hardware and in software must be able to be installed without massive prerequisites, and must be field upgradable.

Other areas which we are just beginning to address involve aspects of function transportability across different operating systems, both statically and dynamically in order to solve load balancing problems as well as the situation where different elements in the system have different function capability. This function transportability must, of course, be transparent to the end user.

Other areas where we must make significant progress over the next few years are in synchronization management, to ensure consistency and validity of functions and especially of data across the DDP environment. Error recovery, detection and control are also areas with significant challanges.

SUMMARY

As we debate the similarities and differences among DCC and DDP systems, it is most important to remember that each of these covers a very broad range of different installations. Individual systems within each category are quite different. These points of difference include degree of complexity, speed of operation, the extent of geographic dispersion, the quantity of co-ordinated control versus peer independence, the stability of the system and the amount of alteration within the system, to name just a few. I believe that there are even greater dissimilarities among systems within each of these categories than the differences between the two categories themselves.

I look forward to further discussion of these points.

Thank you.

DISTRIBUTED DATA PROCESSING SYSTEMS - DISTRIBUTED COMPUTER CONTROL SYSTEMS: SIMILARITIES AND DIFFERENCES

H. U. Steusloff

Institute for Information and Data Processing, Fraunhofer-Gesellschaft, Sebastian-Kneipp-Str. 12-14, D-7500 Karlsruhe 1, Federal Republic of Germany

Abstract. To answer the question contained in the above title we start with the examination of properties of both the "classical" computer systems, the batch computer system and the process control computer system. Distributed computing systems show new properties, introduced by the computing system communication network as well as by novel man-system communication means and methods. These new properties influence both types of distributed computing systems in such a way that their properties become similar and differences degrade to parameter variations.

Keywords. Computer control; distributed computing systems; real-time conditions; communication network; man-system communication.

INTRODUCTION

The preceding introduction to this panel discussion by Harrison (1979) starts with a hypothesis supposing the similarity of requirements in both Distributed Data Processing (DDP) and Distributed Computer Control (DCC). To give reasons for our statement on this hypothesis we will start from a historical point of view to show the changing properties of "classical" computing systems. We will point out new properties, as introduced by communication, and discuss their effects on both types of distributed computing systems. We will use the term "distributed computing system" as a collective name for both the DDP and DCC systems.

PROPERTIES OF CENTRALIZED "CLASSICAL" COMPUTING SYSTEMS

Up to the middle of the 1970's there existed a well established distinction between process computer control systems and data processing systems. This is documented e.g. by IBM's systems 1130 and 1800, both of them basically being equal machines, but the first one being called "computing system" whereas the second system was named "data acquisition and control system". In fact, there were some significant differences in the utilization and consequently in the requirements and properties of both types of systems (Fig. 1).

The classical general purpose data processing system primarily is a batch processing system, characterized by the following properties:

- Computing processes with a great variety of applications.

- Computing processes act on (very) large data sets, the input data being completely present at the beginning of a computing process and all necessary resources being owned by a computing process for the time of its execution.

- The connections between several computing processes are established by the human operator, employing the control features of the operating system and the possibilities of the data base management system.

- Dependent computing processes are run sequentially, the proper sequence being maintained by the operator or by predefined operating system procedures.

- Due to the latter point there are no time-constraints for the execution of the single computing processes in a short-term sense except cost-effectiveness and the availability of the results in a reasonable time distance, whatever "reasonable" may be.

- Short-term availability of the computing system has only little influence on the security of the tasks, performed by the system.

- Data security is of very high importance due to security risks by unauthorized access to data or due to the expenses or even the impossibility of data reconstruction.

- The speed of input and output of data is given by the input/output devices of the computer system.

On the other hand there are powerful central process computer control systems with the following characteristics:

- Several <u>interdependent, event-driven</u> computing tasks, running "in parallel", i.e. simultaneously, and dedicated to plant control,

- influenced by the <u>human operator</u> parametrically, <u>not</u> with respect to the <u>sequence</u>.

- Time conditions are given by the controlled technical process which determines e.g. sampling rates of process data or response times to events (<u>real-time conditions</u>).

- The <u>data sets</u>, processed in real-time, are <u>small</u>. The amount of documentation data may become large, but this data is only collected and hardly processed in real-time.

- The <u>availability</u> and security of a process computer control system highly influence the security of the controlled plant.

- <u>Data security</u> is of minor importance since process data mostly can be reconstructed (e.g. by reading from the process).

- Speed of the <u>input and output of data</u> is given by the plant and control dynamics.

So far, there seem to be much more differences than similarities between both types of systems. But we are still discussing centralized systems.

By reasons which shall not be discussed here, computer communication systems were introduced. From the first steps of coupling two computer systems for data exchange, today we arrive at distributed computing systems. This development is accompanied by a significant progress in methods and equipment for man-computer and man-process communication. Assuming that the earlier mentioned properties are still important, we will have to find out which properties are added by distribution and communication, and how they influence the development of both the "classical" computer systems towards distributed computing systems.

PROPERTIES OF COMMUNICATION SYSTEMS

According to Enslow's working definition (Enslow, 1978), a necessary component of a distributed system is the communication network. Agreeing to this we have to examine the properties of those communication systems. One important property of a communication network has to be its ability to

- react under the influence of <u>real-time conditions</u> and to satisfy them.

This holds, independent from the special communication protocol and the method of communication used (packet-switching, line-switching etc.). Further properties are

- the ability of exchanging data as well as programs between the participants of a network and

- the accessability of all data and programs to all participants of a network.

Data not only means data to be processed but also means data concerning the status of the system. Finally,

- the communication network should be an autonomous subsystem of a distributed computing system, thus adding new resources to it (in the sense of Enslow's cooperative autonomy).

INFLUENCES OF MAN-SYSTEM COMMUNICATION

The progress in display technology opened new possibilities in the communication between man and computing systems. The advantages of

- multiple system-wide access to information by use of the communication system,

- message input and output in natural language or another optimally coded form (in the sense of human engineering) or

- fast information input and output in a problem-oriented way

can only be utilized if there exist

- an information-communication system with real-time properties,

- fast reacting data processing components and

- means for the coordination of the access to information.

PROPERTIES OF DISTRIBUTED COMPUTING SYSTEMS

We will now investigate the influences of computing system communication and man-system communication on both types of distributed computing systems, the DDP and the DCC systems (cf. INTRODUCTION). For this task we use the list of properties as examined for the classical computer systems (Fig. 2).

The <u>variety of applications</u> in DCC-systems now

is great too, due to the novel possibilities and methods of man-process communication (natural language dialog systems, graphic display systems) as well as by dispositive tasks in combination and communication with disposition computers.

Real-time conditions are introduced into DDP-systems by interactive, dialog-oriented man-computer system communication and the requirements of computer communication protocols.

Computing process interaction by use of the communication network is one of the outstanding features of both the DDP and DCC systems. It leads to a higher complexity of the computer system operations and requires automatic methods of action-synchronization. Therefore task sequence control is done automatically according to preprogrammed schedules for both types of systems.

Data sets in DCC-systems grow up due to the increasing complexity of the tasks and of the controlled plants. This also means increased data-sharing by the computing processes. Because of the high complexity, data security and integrity become most important in both types of distributed systems, where the communication network adds new risks. Finally, the data-input/output is fast in DDP-systems too, to cope with the requirements of the mentioned real-time conditions and the grown up amount of data.

Finally the influence of system availability on the utility of the system application is high also for DDP-systems, due to their real-time requirements and the growing interactions between DDP- and DCC-systems.

From these considerations and according to Fig. 2 we come to the conclusion that the hypothesis, as stated in Harrison (1979) - concerning the similarity of DDP- and DCC-systems -, is correct for the dimensions of the problem space considered here. We do not try to apply weighting numbers to Fig. 2, since then we would have to specify certain applications for both system types. However, we want to point out that these similarities even grow, considering the expenditure for the following system components which are equal and equally important for both the DDP- and DCC-systems (Steusloff, 1979):

- Fault-tolerance, error detection and disposable repair time for broken down system components,

- utilization of spare systems resources (function-sharing redundancy) in a preplanned manner to increase availability,

- use of high level, universal languages (HLL) and/or very high level, problem-oriented languages (VHLL, POL) to support an economical application- and system-software production.

REFERENCES

Enslow, P. H. (1978). What is a "Distributed" Data Processing System? Computer (U.S.A.), Jan. 1978, pp. 13-21.

Harrison, T. J. (1979). Distributed Data Processing Systems --- Distributed Computer Control Systems: Similarities and Differences. A Panel Discussion. Proceedings of the IFAC workshop on Distributed Computer Control Systems, Oct. 2-4, 1979. Pergamon Press, New York, to be published.

Steusloff, H. U. (1979). Distributed Data Processing Systems --- Distributed Computer Control Systems: Similarities and Differences. Introductory Remarks for a Panel Discussion. Proceedings of the IFAC workshop on Distributed Computer Control Systems, Oct. 2-4, 1979. Pergamon Press, New York, to be published.

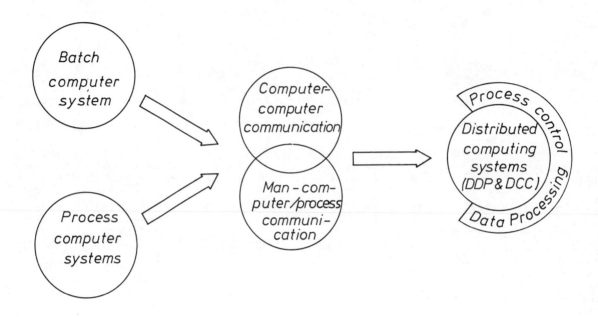

Fig. 1. Change of Computing Systems by Communication

Properties	Classical Computer Systems		Distributed Computing Systems	
	Batch	Process Contr.	DDP	DCC
Variety of applications	great	limited	great	great
Real-time conditions	no	yes	yes	yes
Computing process interaction at runtime	low	high	high	high
Task sequencing	by man	automatic	automatic	automatic
Data sets	large	small	large	large
Data sharing at runtime	little	more	much	much
Data security & integrity	important	less important	most important	most important
Data input/output & transmission speed	not significant	fast	fast	fast
Influence of system availability on task security	low	high	high	high

Fig. 2. Qualitative Properties of Computing Systems

DISCUSSION

Gellie: I think there are two points of view on the question posed for the panel: If you are a computer scientist, you would see the two different choices as being identical; whereas, if you are a control engineer, you certainly see that the two approaches are completely different. For example, to a computer scientist, carrying on from the presentation that Lalive was giving yesterday, there is no difference between, for example, tight and loose coupling, multi-tasking or multi-processing, distributing by logical function or geographic resource. These are things that can be handled within a general operating system in the kernal. The main difference is in the detailed implementation. So, for the control engineer, he sees the differences in the operating environment, the response time, error detection and correction, the structure, and so on, as being important. This raises a problem that I think we have felt in IPW/TC5 and IEC/SC65A/WG6. Although lots of the work is the same, we run into problems because the EDP people are a bigger group and they tend to do most of the work. For example, a language like Fortran is developed which is great for EDP but, as soon as control engineers try to use it for I/O, they realize it as not really the cat's meow! Or, HDLC is fine until control engineers want to use multi-mastership, and they find that HDLC doesn't really have the ability to support transfer of mastership.

So I really feel that, from a computer expert point of view, DCC and DDP are the same and, if everything is done in a general way, the same tools, languages, and so on can be used by both the distributed data processing and the distributed control system people to design their systems. It's only in the actual implementation that the engineer needs to worry about specific "nuts and bolts" issues that are actually so important, as Fred Sheane, for example, pointed out.

But the real need is that the automatic control community should somehow get together with the EDP people, so that when the EDP people are working in areas like communication protocols or languages, they understand that they are working in a completely general sense and that they are including things that are necessary for DCCS, but not necessarily apparent as being immediately useful in EDP or DDP, like I/O in supporting languages and multi-mastership in communications.

Harrison: Do you have a specific suggestion in terms of how the control people can influence or affect the work of the EDP community?

Gellie: Well, in one instance at the moment, we are trying to set up some relationship with the ANSI people and the ISO people on HDLC to see how they view the multiple mastership transfer problem. We think what we require to meet our functional requirements in control systems can be harmoniously incorporated into the HDLC standards. The same sort of problem existed with Fortran and a solution was finally found, except I would hope that in the future things can be done at an earlier stage so that it's a more harmonious inclusion rather than sort of a bubble that's stuck on the side like the extensions to the Fortran were.

Singer: I think that the example that you gave was an excellent example of what I would call "convergence." I believe for data communication within distributed data processing, we have the need to manage the capability to transfer mastership. It was not thought out on day one when we started to discuss and identify protocol, so it didn't get in. The best you can say is that we didn't see far enough ahead as to what the proper domain for protocols was. We in data processing have to go back to correct that, and to enhance and extend the communication protocols to be able to handle the kinds of problems you are talking about. Some of the problems exist in data processing also.

Gellie: I think that that would be a marvelous opportunity for the control community to go back to the EDP guys and say, "Look, you've discovered the need that we could have told you about before. Now you realize the need for the two of us to work together in the future." I think there is a feeling amongst the control community that we are the forgotten child, and if somehow we could recognize ourselves as being brought back into the fold and recognized as a legitimate, important part of the whole computer world, that would just be tremendous!

Singer: I don't know who is in control of the system that's keeping you out. I can't speak for all distributing data processing, but it seems to me that it's an open community and control people just have to speak up and be heard.

Gellie: Well, certainly you might recognize that the control community can't go to a mainframe vendor and say, "This is what we need and why don't you include us?" Because they say control is such a small part of our market and, if it's going to cost any money at all, we are not really that interested. Compared with the whole EDP market, the control community is a pretty small fish; it's the same with the semiconductor houses too, of course.

Born: I think it may be a small fish, but it's growing and it has a big appetite; there is a big market out there as evidenced by this workshop and the applications we've heard. I think one of our big hopes in the control area lies with the IC manufacturer, such as Intel and Motorola, in that they will probably provide us with more of the types of LSI components that will be useful in solving our problem. I really don't see that much hope for the control people working closely with the data processing people; that would be nice, but I don't think it's likely to happen.

Harrison: Are you getting the necessary systems support along with those LSI components?

Born: Well, we are not getting the components yet. As Gellie pointed out, the HDLC chips that are available don't really solve our problems, but maybe we can influence future chips that will solve our problems.

Steusloff: I very much agree with you, although I wouldn't say that for the DCCS people just to join the DDP people will automatically solve our problem. Let me give an example. Yesterday, I told you about the problems of synchronizing access to a distributed data base which consists of multiple copies. There is a solution for this problem in the DDP area, but it's too complicated to use in real time conditions. Now, if Intel provides some hardware to cope with these real time conditions, then the transfer would be good, but it's not so easy to take the methods and transfer them. On the DCC side, as I told you yesterday, we have to provide for some very simple, and maybe dirty solutions, just to make it work. It would be nice to get the clean solutions and methods from the DDP side.

Lalive: I want to add some comments to what Warren Gellie said. I very much agree from the systems point of view: There is absolutely no difference, but we can also try to look at the problem top down, from the applications point of view. I don't see why there is any reason just to make this distinction only between these two groups. We within our company have the problem of telling the people who make the control systems for power generation to use the same methods as those who make the system for power distribution. They think they have totally different problems and then we have to come to the systems people to tell them they can do most of their work together. We have applications, especially in the power distribution, where I could not say it's either data processing or control because it's all intermixed. If we handle a data set incorrectly, a switch may blow, but also at the end of the month the payment from one company to the other will not be correct. So, it is really a problem which involves both DDP and DCC;, there is no distinct limit between one and the other. What we really have are common system aspects, but, of course, many different applications. If we can feel free to group some in a group which we call DCC, then the other group may be DDP. But, in my opinion, we could as well make many other groups. There are, of course, different applications, or perhaps class of applications, with some common properties but, in my opinion, it's not relevant from the systems point of view.

Musstopf: I think this discussion leads directly to a more general question. The title of this discussion is "Similarities and Differences Between Data Processing and the Computer Control." As I understand the discussion here, we are mainly discussing more or less general purpose computers as parts of the distributed systems. If you analyze the present projects in the industry, you will see a lot of single boards in general, as well as custom boards. I think it must be asked here if the title "distributed computer control" is meant seriously or not. Therefore, my question is: should we include here the so-called dedicated hardware?

Let me give an example of this. If you had to measure the length of something in a process, you could use any traditional equipment, but you can also use TV cameras, with pattern recognition. For this purpose, from the theoretical point of view, you can use large scale computers or minicomputers; but, they both are too expensive and too slow. Therefore, at the present time, we are looking for dedicated hardware, in the sense of parallel processing, and very special processors. The disadvantage of the dedicated hardware on one side is that we had to design, implement and test only on a very, very low level--in some cases, on the microcode level. The big advantage of this is, in this case, that if you are using dedicated hardware, the chance to reuse such subsystems is growing and reusage of the subsystem means the reusage of hardware and software, including software in ROM.

Harrison: When you say "specialized hardware," are you referring to hardware which is specifically designed to do a particular task or are you referring to a microprocessor which is coded to do a specific task, even though the microprocessor itself could be considered general purpose?

Musstopf: I am referring to both. For both cases, the main thing is that we consider real time applications, in the sense of per Brinch-Hansen in the "Communications of the ACM" from November last year. The main thing is you can use, for example, boards like the single SBS from Intel or the MDX or SDB from Mostek. You specialize it to your part of the application.

Harrison: I think that's a good question for the panel. Should we include this dedicated hardware as a part of the distributed computer control system, or does that become, if you like, an instrument and, therefore, beyond the process I/O interface?

Johnson: I could take your question one level higher and ask: Should this differentiation extend to the network protocols, for instance? I think that there is a bonafide difference in viewpoint here; depending on if you view the problem as an applications user or if you view it as a computer scientist. The viewpoints differ and I think if you ask, "What's the best solution to a given application?" you may not implement that most efficiently, in terms of software and hardware modules that you can take off the shelf; although you may be able to do it in any given situation. In research at the applications level, one usually asks what is the best solution to the problem without constraints first, and then asks what I have to do to implement it in the presence of the constraints.

Sheane: I'd say it's absolutely essential that the elements that Dr. Musstopf is mentioning are part of the network, if there is any dependence, higher level control, or an operator exercising the control. It is absolutely essential that those elements are part of the distributed system.

Harrison: Then let me ask if that comment also applies to the traditional PID controller, which is, in fact, a special purpose analog computer?

Sheane: As long as it influences the operation of the process that you are attempting to observe or regulate, then it must, to my mind, be part of the overall system.

Harrison: So we've had distributed systems for a long time!

Sheane: True.

Willard: First of all, let me point out that I am not a computer scientist-- I am a computer engineer, and that changes my thinking quite a bit. I do agree that there is a lot of similarity between distributed data processing and distributed computer control systems and that the differences are really parametric, if you will, rather than structural. However, bear in mind that most computer control systems that are being implemented today have a link, generally from the highest level of the control system, to an EDP machine somewhere. The computer control system is part of a DDP system in most applications. But I would like to take issue with Enslow's definition that we seem to be working from. Enslow makes a point that you have to have a multiplicity of assignable resources. In distributed computer control, generally most of the resources are fixed or semifixed. If a box is driving a valve, no other box can generally do that because there is only one connection. I think there are similar problems in most DDP applications. The purpose of the high level operating system that Enslow discusses is primarily to perform the dynamic assignment of tasks to processors and that's not really necessary. In fact, it serves no purpose if the majority of the resources are preallocated anyway. The third point of his definition that I disagree with has to do with system transparency. I think there is a fundamental difference in response time and in response integrity, depending upon whether the response is coming from the processor that's immediately adjacent to the processee, versus whether it's somewhere on the other side of a communications network.

Harrison: I think that we should point out that Philip Enslow in his article also says that, to his knowledge, no system today satisfies his proposed definition. He points out that we are making progress toward such a distributed system but we are not very far along.

Willard: Let me point out that there is a reason why we are not making much progress and probably won't. That is the point that Rubin Singer made earlier, that most businesses are hierarchical in nature. They would much rather have a system that is hierarchical, rather than a group of equal peer processors that can all do the same job.

Wilhelm: I would just like to agree with the last comment and place a little bit more emphasis on it. It seems to me that Enslow has outlined, and Miller paraphrased and supported these definitions, as requirements of a distributed system. I think that some of them in no sense can be viewed as requirements; they can certainly be viewed as desirable in some instances, and may even be viewed as requirements for certain applications. For example, there might be certain applications where online reconfigurable processing would be a requirement, but I think it's unnecessary to view that as a requirement in order to have a truly distributed system. There certainly are many distributed control system applications where that is not a requirement. I think calling that a requirement distorts our view of what we are after. Again, the differences between applications should be lodged in the application software. Therefore, there may be bonafide differences in terms of application protocols from one system to another that make distributed control systems look very different from distributed data processing systems. It may well be that you can share a lot of the communication protocols which are at a lower level. There are distinct differences, between what is required at the application level in the way of protocol for communication from one application package to another and what is required at the level of the hardware and the communication software.

Harrison: I'm not sure that Warren Gellie would agree with your comment on the ability to share protocol at the lower level, with the differences becoming more apparent at the application of higher levels. Isn't your concern for integrity and error correction at the very lowest level of protocol?

Wilhelm: As far as error correction is concerned, I think I heard a comment earlier in this morning's session that error correction was the responsibility of the communication protocols. I think that may be a little bit of a distortion, in the sense that, the heart of the system, which knows what best to do as a result of an error, is an applications package: so it may be necessary for the communications part of the system to recognize that there has been an error. However, I think it would be illogical to assume that that part of the system should decide what to do as a result of that error. The application packages in a control system should decide whether they can do some reconfiguration or take an alternate preprogrammed path within the particular application package to recover from that error. It seems to me that there are responsibilities at each level for error detection and correction. It's, more or less, error detection in the communication part of the system, and the decision about what to do as a result of that error should be lodged in the applications package.

Gellie: How can I argue? Certainly, one of the basic functional requirements for Proway is that all errors detected should be passed up to higher levels. In many cases, you are going to have a situation where there is some automatic recovery which can be pursued. In the ultimate limit, however, it depends on the application package.

Miller: I would like to clarify my point of view in that regard. Whereas some errors may require the passage of the error to the application level for resolution, I believe a fundamental system requirement is that, wherever possible, errors should be handled at the lowest possible level. If you can handle an error by retransmission in a lower level or by using an alternate path, you should definitely do that at the appropriate low level and not automatically force the treatment of all errors at the application level.

Gellie: There is still an implication that this satisfies the application if that can be automatically done at a lower level. Obviously, you want to push them down to the lowest level possible, but there is still an implication that that can be done automatically at the lower level because it is consistent with the application.

Miller: I guess it's a question of what functions are delegated to what level. It's the aspect of whether or not, with the delegation you assume, the function will be performed correctly according to the requirements, or whether you will be told that it was not performed correctly. So it is really a matter of defining what is the function to be performed at that level.

Singer: Some of you may not have appreciated the source of Tom Harrison's earlier comment about big programs. Within IBM, I am one of the people who always ends up with these very, very large programs. Whereas other people seem to be able to get by with 10, 15 or 20 thousands lines of code, and mine end up to be hundreds of thousands of lines! It seems to me that this point on error detection and correction is exactly the point that always drives us to larger and larger systems. I think, Bill (Miller), that initially you said it should be at a lower level. The answer, as I understand it, is that error detection and correction must occur at the appropriate level. Now some times that appropriate level is really very, very close to the primitive hardware that detects the error. An example that we have had for years is the reading of magnetic tapes. There are all kinds of methods, of not only doing error detection on magnetic tape recording and reading, but also of correction. There are ways within the communications network to do the same thing and, where in the communications network you can, in fact, detect and correct the error, it should be done there. In other places, the only place that you can determine what action to appropriately take when an error has occurred is back at the application level. In these cases you must pass the information back so that the application can properly cope with the recovery or alternatives you want to take when the error exists. So, the appropriate level is absolutely required, and that's one of the things that continues to drive us to larger and larger support packages that build in more and more error detection and alternative recovery procedures.

Steusloff: I would add to Mr. Singer's comment. You are quite right that the error handling has to be done on different levels of the system. To cope with this large amount of code you mentioned, we feel it is necessary to structure the application language so that error handling and standard functions are separated, so that the programmer is able to handle this large amount of code. I think that's the way to do it. So we have several levels of error handling capabilities, and that fulfills the requirements.

Harrison: Would it be fair to say that one of the considerations of where you handle errors is an economic one? Referring back to the tape example, if I remember early tape units and the characteristics of early tape, errors were quite frequent and they generally were reading errors and not solid errors on the tape. To handle all of those at the system program level would have been a significant burden from a performance point of view. From a circuits point of view, it was economical enough that one could put the circuits in the tape unit itself. So, economics plays a role as to where and how errors may be handled.

Humphrey: I think that this is an interesting conversation that we are having here. One of the things that is happening, truly, is we are trying to define what a distributed system is, in both computer control and data processing. As my colleague from Bolt-Beranek & Newman said once upon a time, one of the things the people in the control system business want to do is to find out what useful things they can "rip off" from the people in the data processing business. So, one of the things that I think is useful is the definition of a distributed system such as that given by Enslow. However, I would like to see an expanded definition that takes into account the fact that control systems have a higher priority communication system than is involved in the computer system; it's the process itself. It has the error checking and all those kinds of nice things at a much more fundamental level than in the kinds of communication network that we are talking about. There is a response to errors that is necessary for distributed computer systems that I don't think is really at the same level in distributed processing. I thought Dr. Johnson did a very nice job in delineating some of that.

One of the other things that we get confused about is the necessity for expandability in distributed computer control systems. We confuse that with felxibility of movement of tasks and assignment of tasks in a distributed data processing environment. For the distributed computer control system, the critical element is expandability rather than moving tasks around at will, especially when we take into account Dr. Musstopf's comment about hardware being intelligent.

Harrison: Would you define "expandability" as you are using it?

Humphrey: By "expandability" I mean that a computer control system grows from year to year. We have physical expansion with associated logical expansion; when we add more tasks, we add more control points. That's a continuing need. One of the things that I would have liked to have seen in Dr. Steusloff's paper yesterday was the control loop that went all the way back to the beginning, because any computer control system has to do that.

Miller: I would like to ask Rusty (Humphrey) a question to clarify what he means by "expandability." Do you mean that the system must be expandable with minimal impact to existing application programs? Do you mean that the system, when it is expanded, must be shut down and totally re-sysgened and all the application software rewritten? What are the required impacts associated with expandability? In other words, do you want the system to be on-line reconfigurable expandable? How would you like to accomplish this expandability?

Humphrey: Now we are talking design objectives. My personal design requirement clearly would be that on-line reconfigurability is not a problem for my specific application, because I can shut down a control system in ISABELLE for some short period of time, like a couple of hours. I would not like to have to put the manpower investment into rewriting all my software and, in fact, I think accelerator control systems don't do that anymore. At one time that was true, but it is no longer true. I think that it is possible to reconfigure systems in some short amount of time and I think that the technology is here now that we can do that without having major impact on software. By major impact, I mean investing an incredible amount of time trying to rewrite all this stuff over again, although some minimal impact is acceptable; I can live with a couple of man-weeks or man-months, but I can't live with man-centuries!

Miller: I would ask the Workshop also to consider the question of backup and automatic failure recovery in the same realm as on-line reconfiguration. In other words, I think there are different aspects to the problem. Having a certain amount of redundancy associated with the system, which is required for reliability and which can be dynamically utilized, in my opinion, has a great deal of flavor of on-line reconfiguration. This is different from saying you want to add new and distinctly additional features or capabilities.

Buckner: I have a comment in tying together many of the things that have been said, particularly what Rusty Humphrey was just saying about considerations of the process itself. It seems to me that this is the core of the differences between distributed data processing and distributed computer control systems. I'm less knowledgable about distributed data processing than computer control, but the way I tend to view it is that, when you have a distributed computer control system and you are looking at the performance of the system, be it multi-faceted or whatever, you are looking at the performance of the process as being of paramount importance. That's, of course, because you justify return-on-investment on how the process is going to perform over a specified number of years, considering availability of the computer control system, and so on. That's why you assign error correction, error detection, fault containment, fault tolerance functions at different levels of your system. In particular, you are trying to maximize, in some sense, the performance of the process, and not fault containment, fault tolerance functions at different levels of your system. In particular, you are trying to maximize, in some sense, the performance of the process, and not the control system. I see that as being a much more nebulus thing in the data processing area where you may be concerned with things like response time and availability. But, it is much less well defined and that impinges directly on the design and where you place your appropriate functions. The key to distributed control and how you design it, where you put the different functions, is not precisely dependent on the controller itself. Any way that you get the most from your process is the essential aspect, in my view.

Harrison: I would like to ask Rubin (Singer) to respond to that in terms of: Is availability a highly sought after consideration in data processing? Also, could you comment on the question: Are data processing systems continually being expanded physically or through added functions?

Singer: As you were talking, Mark (Buchner), I was thinking of when I was in Chicago. There are a number of rather large retail stores that had billing systems. There was one clear hold-out of billing systems, and the attitude there was specifically what you described to be the criteria by which you would evaluate the computer control system; that is, in terms of pay out for performance in that institution. It was deemed not warranted to install a computer system, although the store was very comparable to many of its peers who were making tremendous investments in billing systems. So, I think that the criteria is applied, maybe not as precisely all the time, in either field, but it is the same criteria in computer systems, centralized or decentralized, as well as in computer control.

Buchner: I recognize that, but the type of payback, in my view, is somewhat different. A payback in the data processing environment, such as a billing system that you are referring to, tends to be an all-or-none type of payback. I don't see the measure of performance varying in a continuous way with respect to the process that you are controlling. You have a billing system and, fundamentally, it is either working or not; either you are issuing bills of a million dollars for the purchase of ten cents bubble gum, or you are not. Once you figure in availability, of course, it's a different story, but aside from availability and reliability considerations, the performance is of a significant different character than it is for process systems where you have continuously variable performance measure.

Harrison: Couldn't one say that a measure of an accounting system or a billing system is, for example, the cost of processing one billing transaction? That's a measurable variable which you can talk about; how many bills you sent out this month for a certain cost.

Buchner: That's precisely my point. In a distributed data processing environment, the primary profit is really to be had in the computer system itself, as opposed to payback of a control system, where the profit or loss to be gained is in the process itself.

Singer: I see it as distinctly similar. It is not how many bills you send out; it is actually your cash flow, and how promptly the bills are sent to the customers and how accounts receivable end up. It is measurable, not in a absolute "yes" or "no" sense, but in the same kind of percent improvement of yield that you would see in the process.

Let me talk about availability briefly. I see a big difference as you focus on what I would consider a very structured process, as opposed to what you are presuming is an unstructured environment in distributed data processing. In my experience in distributed data processing, both types of systems exist: highly structured systems and relatively unstructured systems. The unstructured systems do indeed appear to be very different and often appear not to require extensive reliability or availability, whereas structured DDP systems do, in fact, require the same kind of responsiveness and availability that those of you in control systems know has to be there. Twice this morning, statements were made about how, if a terminal goes out, you can obviously get another terminal, but you can't afford to lose a sensor in the process. I suspect that, in control systems, you build in multiple sensors so you have backup, and you build in the kind of availability you need. So, if you do lose sensor, you have enough similar or backup sensors that you can manage to keep track of what's happening. We do the same thing with terminals. The reason it doesn't appear to be so bad if a bank teller's terminal goes out is because we designed the system so you have six terminals there together. If one goes out, you can get an adjacent one. Just two weeks ago, I had an "opportunity" to work on on a Saturday and the terminal I was using did not function, and there wasn't another terminal available. I'll tell you my "process" did blow! It was a disaster because I couldn't get work done. So, it was extremely similar, in my mind; except from my vantage point when we put in terminals, we cluster them so we always have an available one, as you do when you put sensors in a process, so that we have backup and a method of keeping track of what's happening. So, it isn't an "all-or-nothing" kind of availability.

Sheane: I wish I were in a position to invite Rubin (Singer) to the refinery room; we work

Saturday and Sunday, at night! Bob Willard, I think made a particular point that we can all describe "transparency," but the mechanism can be considerably different. Rusty (Humphrey) made some comments along the lines of a definiton for the distributed control system. We developed a glossary in IEC/SC65A/WG6; I've been to seven international meetings trying to achieve agreement on just the terms in the glossary! I think it becomes clear, to the people who are trying to wrestle with these problems, that there is ultimately a mechanism that satisfies the one application requirement that would be unsatisfactory for the other application requirement. That, I think, is the difference we are wrestling with. You talk about the decompositon of the system, and you probably can decompose the system. But, think of it in terms of failure analysis and try to identify whether or not, in fact, you are still in control. Rubin Singer's comment on multiple sensors is well taken; we do have multiple sensors, but we haven't found a way to have multiple outputs that are controlling the material or energy. Quite often it is the incorrect adjustment of the output action, with the consequences of that action taking place in seconds, and there is no time for either equipment or man to make any corrective action.

Musstopf: Let me come back to the terminal example given by Rubin Singer. I would like to change those examples a little bit: The terminal is not broken or not available, but is only working slower. In this case you are arguing with the hardware manufacturer, for example, or the manager of the service center. But going to another example of an on-board computer on a plane, if the computer is working slower during landing, it is much more critical than the slowdown of your terminal. The main objective of the design of the system itself brings me back to the dedicated systems that you need to guarantee such reaction time, for example. This is the most critical part in many control system applications and it leads directly to a so-called integrated hardware-software design of such systems.

Gellie: Two separate points: firstly, getting to what Mark (Buchner) was saying before; is your question really that in the DDP the economic justification is on production cost benefits, whereas in control, it's not only production cost, but also product quality? Maybe that's a basic difference. I ask this to Rubin (Singer). Is product quality really something that is measureable and used in the equation for cost benefit?

The second comment was to Fred Sheane about the outputs. I refer back to what I was saying yesterday about what I called performance parameter control, rather than engineering variable controls. Can one include control of performance rather than just control of the variables, so that automatically the remote units can be controlling the performance that you require and not just the specific valve? For example, you say you want to reduce flow and this approach automatically does corrective action if it sees that a particular output is not driving to achieve the result and then it readjusts other parameters in the system. Is this done now, is it possible to do it, or are time restraints always going to prevent it?

Buchner: I'd like to comment on what you said Warren (Gellie) and also thank Rubin (Singer) for bringing up an example which keyed several other examples in my mind. I think it's rather unfortunate that we started this discussion by breaking up the entire area of, what I would call, "distributed intelligent systems" into distributed data processing and distributed computer control. I think Tim Johnson had made a very, very good point in the beginning and that's how to view the two together, both of them really as being very close to one another, except for the fact that the plant is somewhat different. I can think of three examples which will bring out my point -- one being a banking system, perhaps not working on weekends but during the regular weekday working hours, in which there are several terminals with which the tellers are servicing customers and a particular terminal breaks down. Well, it's fairly clear in that instance that there is a direct payoff or a direct loss, in that customers will not be able to be served as rapidly. There may be some inconvenience that is caused as a result of this, maybe considerable inconvenience to some people who need to get money and otherwise who are going to be overdrawn at the bank. But, certainly, it's not an extremely critical circumstance.

I think of another circumstance that one might classify as being distributed data processing: In a hospital environment where one has a distributed data base and there is a patient who needs medicine of a particular type. The doctor is trying to find out what is the best possible medication for that particular patient. This probably would be classified as data processing and yet it is very, very critical that the doctor receives the proper information and there is tremendous payoff or loss from the operation of the distributed intelligence system. Finally, a third example will be perhaps a nuclear reactor, where you might be talking about proper control, otherwise the thing might blow up and there control is very critical. The point is that we shouldn't be looking at data processing or control; but really what we should be looking at in terms of a unifying framework are some aspects of real time properties of a system and the need for response time of a system. There are clearly systems that are very critical with respect to response time, others that are much more forgiving, and others in fact that are not that much concerned for real time at all, such as computational facilities when one is looking at trying to solve the four color problem. That gives you the spectrum, so I don't think it's appropriate at this point to talk about the extreme, but to talk about a unifying framework in terms of real time requirements of a system and I think that tends to unify things. I see Rubin (Singer) shaking his head.

Singer: The reason I was shaking my head was that I really want to make a comment. I think the point that Günter (Musstopf) was making is relevant here. As he was talking, it seemed to me that distributed data processing and distributed control systems may not even have the luxury of deciding whether they are different, because we are using the same pieces. The building blocks that both of these groups are using turn out to be the same, so that I'm not sure but what the discusssion has a degree of muteness, simply because we are using the same components. It comes back to your question, Mark (Buchner). We are using the same components. The last real effort I was concerned with, I guess, was a couple of years ago but it is continuing because we have not been able to solve it, turns out to be a real time response problem because of the volume of work that's needed in Japanese banking. In Japanese banking we in IBM are unable to aggregate the amount of computer power required to respond to the demand of the Japanese banking transactions. Therefore, we have to have alternative solutions. Japanese banking exceeded our requirements in nuclear reactors and the ability to manage and monitor our space flights. We seem to be able to get together the computer power to handle those things. Now, clearly, there are undoubtedly going to be process applications we haven't seen yet or gotten to and they will be bigger, but the biggest one I have run into has been the one we casually talk about as "banking."

Harrison: Rubin (Singer), I agree with your comment that we use the same pieces, but would you agree or disagree with the statement that the number of different ways of putting the pieces together is probably greater in solving the various industrial problems than it is in solving data processing problems? A specific example we are both familiar with is Series/1 which is, in a sense, a bag of parts. If you want one disk file, you can have it, and if you want 13 you can theoretically have that too. Whereas, the IBM 8100, being oriented more towards data processing, seems to have fewer options in terms of its configurability.

Singer: I feel that what you are probing is how many, and how dissimilar, are the various pieces that you find in a typical DCC kind of installation.

Harrison: What I am probing is a question on hardware and software design that says in order to design, let's say a general purpose mini-computer used in many things as well as industrial control, the designers have sought to allow great variability in configuration, virtually infinite. If there are 100 pieces, then there are 100 factorial ways of putting it together, almost. Whereas in data processing, the designers tend to have in mind somewhat more fixed configurations that they are optimizing. They optimize towards a configuration that has lots of files, for example, or lots of terminals.

Singer: I think that distinction may be only one of degree and the degree is that, as you try to be able to handle more and more different kinds of environments, you end up coming up with a "standard interface." That standard interface is generally less efficient for specific classes of interfaces, but you pay that penalty and you end up with a general interface that you have more people plugging into, more pieces, more devices plugging into it. So I think that you do have the appearance of these different things but I think it's achieved primarily at the compromise of a more general interface.

Sheane: May I comment on this line of discussion and then reply to Warren (Gellie). I think if we are discussing higher levels of control computers in a distributed system, the comments Rubin (Singer) made are definitely true. I don't see them to be true in the lower levels, in what we've come to call level one, where we take the piece of equipment and physically locate it in a hostile environment. The mechanisms of the hardware and software could be definitely different at that level. I think that that is a distinction.

Trying to reply to Dr. Gellie's question, the ISA papers and Chemical Engineering Progress, the magazines that I read, and other references tell me that many people are using calculated variables now in supervisory control configurations. There is definitely an opportunity to deal with the profit motive of a process, as opposed to the control of the level, pressure, temperature, and those typical variables which we more commonly regulate. At the same time, we have to keep in mind that there is an ultimate action regarding the control valve on the process, and it's that ultimate action that still is dependent on the integrity of the total system; even if the commands are coming from a much higher level system through the network, the control valve still has an ultimate command and that's particularly important. There is such a thing as negative payout; we've all been talking about profit potential here but having a system installed that permits dangerous action or incorrect control can have a very real negative payout. That, I think, is what we attempt to consider when we analyze the mechanisms that we are talking about here.

Johnson: I just wanted to raise a question that might shed a little bit more light on the topic and to ask, whether, if we went up to the operating system level, how much satisfaction is there in this group with the operating systems which are supplied by vendors, and how much need has been found to modify those operating systems for specific purposes of a control application?

Merluzzi: I just wanted to make a couple of comments from the point of view of a user and identifying with Mr. Sheane. The fact of the matter is that we are trying to compare data processing and control aspects of computers and it's obvious that, within the control community, we haven't completely defined what we mean by a distributed process control system. This is never more evident than if you went out to write a bid spec or a request-for-quotation among the existing vendors of so-called distributed control systems. They vary widely in function, scope and redundancy. I think this totally reflects that it is an applications-oriented process because, if you look at, and analyze how these vendors came about designing their systems, they all have a major corner of some market in the industrial field, whether it's chemical, or petrochemical, or steel or whatever. Their systems totally reflect the needs of their customers and so their distributed system reflects those needs. Thus, I tend to agree that it's an applications question from a control point of view, and I don't know if you can get a general distributed control system that will satisfy all the application needs.

Musstopf: Let's comment on the question of Tim Johnson on operating systems. I think in this area of operating systems for control systems that there is a diffemece between general purpose DCC and dedicated DCC, because, for general purpose use, operating systems are available. One of the characteristics of such operating systems similar to large scale systems is you don't know the exact reaction time for an interrupt, for example. For the dedicated DCC systems, in the sense of a single board system like the SPC, you can immediately say the reaction time will be 100 microseconds or 150 and so on. There is no operating system, only a few modules or real time executive parts, nothing else. That's the only chance to say, I guarantee a reaction time of, let's say, 100 microseconds.

Legrow: I'm pretty well convinced from what I've heard that there are a great deal of similar points between the two systems, but there is one that Fred (Sheane) has alluded to on several occasions. To restate it, it is that the control system is much less forgiving in its tolerance for a graceful degradation. There is no graceful degradation in most processes; if you miss, just once, you are in trouble. You can withstand a little longer downtime, in terms of seconds response time, in EDP and DDP systems. Referring to Rubin's (Singer) situation where his terminal went down, you are more mobile, as an individual, than a process is. I cannot move a paper machine to another building where I have another terminal. That's an extreme point, but it has to do with redundancy, it has to do with mobility of people in DDP systems, and it has to do with missing deadlines. If I'm running a DDC system, for example, which is distributed, and I need to calculate an output every second, I cannot miss one now and then. Rusty (Humphrey), for example, would be in great trouble if he missed one out of every 1000 calculations. So there is no graceful degradation in control.

Steusloff: A reply to your comment. Our experience is that we have a lot of possibilities for graceful degradation in processes. It depends very much on the applications again; if you have a very fast reacting process, you may run into some trouble. If you do not have such a very fast reacting process, you may lose control quality by graceful degradation, for example, but the process is controlled and, for that time until the maintenance comes, this low control quality may be tolerable. On the other hand, in DDP systems you can come into really critical situations. Let's consider the data security as I mentioned further on, when you lose the integrity of a data base, there is no graceful degradation; in 500 milliseconds, a data base may be completely poisoned and you can't use it again; if you haven't secured it in a good way, you may not be able to reconstruct it. It's the same quality of problems in both parts.

Singer: It seems to me that it appears to be different, but in both cases you build in a method so that, when you approach a crash, you have a reasonable procedure. I was trying to think of what's a good example; I think it was a couple of months or so ago that I was in the UK and the data processing people in the PTT went on strike for some number of days. At that time they hadn't solved the strike and the prediction was, if they solved it the next day, they would get the bills straightened out by September of 1980. That's an example of a graceful degradation system that was not working! I think you can find, while the appearances look different in process systems, that they do stop and you build in procedures in a process where you clear out the pipes, so that if it's cement that's going through the pipes it doesn't end up being solidified. You build these things into whatever system you are doing. The appearances are different, sometimes it's continuous flow and other times it's more discrete, but I think all systems have to be designed so that you have a method of closing down, of recovering, of alternative backup, etc. They appear to be different and I'm not certain they are identical, but they do seem to have the same kind of requirement to think out what alternative courses of action are available and which you will choose.

Caro: It seems to me after listening to all of these discussions, that it's possible to contrive examples in any particular area that illustrate any particular point. So, I don't know if we can actually say that there are discrete differences. I guess I've heard one, that Fred Sheane brought up, and that I'd like to amplify, where there is a definite difference. Perhaps somebody from distributed data processing can contrive an example that proves me wrong, but essentially there is a difference in need to locate intelligence for control close to the physical sensing point for the purpose of saving wiring expenses, explicitly, and perhaps in some cases like pipe line systems, it's totally impractical to wire all of the sensing hundreds of miles back to some common control points.

I would like to also emphasize that everything we've talked about is a matter of degree, a matter of timing, particularly, and a matter of importance. I think if we introduce the distributed military command and control system, we'll find that they are measuring times in microseconds for control, of say, rocket thrust vector; in process control we are measuring it in terms of milliseconds or seconds. I'm sure we get examples in any area; however, that can be contrived to be in the other area also. So as far as differences, I would like to offer that it is a matter of degree, a matter of the amount of importance, the amount of loss, the amount of inconvenience, and the amount of dollars that we attach to the particular problem that we are trying to correct.

I think that there are mechanisms from every application area that need to be transported to the other areas that are of value. I know, for example, the mechanisms which are provided in distributed data processing activities for back-tracking and correcting data base errors, just don't exist within the process control area, and they should! Also, for example, in process control real time responses, there are things we do in terms of multi-tasking that don't exist in the distributed data processing environment, but I've heard people in that environment wish they did. So we each have a lot to contribute.

Harrison: I would hate to be accused of contriving an example, but may I point out that point-of-sales terminals are an example where intelligence is being placed out board, for both the question of integrity and also a question of the bandwidth of the connection between that terminal and the central data processing machine.

Sheane: I think that the degree of difference or similarity can be defined. Referring to a three tiered hierarchical system which we agree has the functional elements of a distributed computer control system, we have, in fact, at the highest level, truly a distributed data processing system. There are no differences at the highest level--a management information system is, in fact, and can be, a distributed data processing system. As we come to the next level of supervisory control, there are fewer likenesses, but the greatest number of similarities. At the lowest level, there are the least number of similarities between a data processing environment and a control system environment, in terms of both the hardware and the software.

Merluzzi: I'd like to agree with Mr. Sheane's comment about the highest level of a control hierarchy being similar to a data processing system. But I don't think that necessarily has anything to do with whether it's a distributed system or not. His is a very general comment.

Harrison: I would not be so presumptuous to think that I could summarize the last two hours. I would point out, perhaps, three areas that we seem to be coming back to. The first is that the differences we have been talking about today seem to be more related to the application than to any kind of artificial categorization of something being data processing versus something being computer control. So, it seems to be the applications that are defining the differences and the similarities.

Secondly, I think it was very important that we spent a few minutes on the idea of characterizing the process. The process in the industrial sense is a physical piece of hardware which has to be monitored and controlled; the process in the case of data processing is perhaps, two things: one is the person dealing with the system, trying to get service from it. Secondly, the process is also the enterprise management system with which we are attempting to, in fact, control a commercial enterprise such as a corporation or a store or a hospital. The characteristics of the process seemed to be an important aspect of our discussion. For example, are people more tolerant of a degraded system, or is the process? We held a symposium in Denver on human factors a number of months ago and a classic photograph was shown. It's a picture of a data entry device in a manufacturing plant and on the front of that box, there is a footprint of a big work boot. So, the question of whether or not humans will accept graceful degradation, gracefully, has to be considered as well.

Third, I think Dick Caro's comments on things being a matter of degree clearly has been demonstrated here today. We have talked about response times, for example, in both data processing and in process control computer control, which can vary over orders of magnitude.

I believe that we have accomplished some of what we set out to do with this panel discussion. It has been an extremely good discussion and I want to thank all of you and, in particular the panel, for your participation.

A MULTI-PROCESSOR APPROACH FOR THE AUTOMATION OF QUALITY CONTROL IN AN OVERALL PRODUCTION CONTROL SYSTEM

K. Zwoll, K. D. Müller and J. Schmidt

Central Laboratory of Electronics, Kernforschungsanlage Jülich GmbH, Jülich, Federal Republic of Germany

Abstract. A modular standardized DNC-System is described which is based on the CAMAC Standard. Real Time FORTRAN is used to allow easy adaption of the system to individual production requirements. In a second step the Operating Data Acquisition is added. Quality control is a substantial part of the total production costs of the plant, where fuel controllers for aircraft jet engines are manufactured. The automation of the quality control with its test rigs is achieved by a Multi-Micro Computer System, thus forming an overall Production Control System.

Keywords: Automatic testing; centralized plant control; computer hardware; computer interface; computer software; direct digital control; microprocessor; production control; standardization

INTRODUCTION

The rising wages in the field of industrial production, the demands for greater accuracy, with the use of expensive manufacturing equipment and the flexible and optimal usage of machines have made it necessary to aim at a yet higher level of automation. While in the past decades mechanisation played a dominant role in mass production, flexibility was limited. At present and in the future the emphasis lies very often on application specific products. In this connection, it is essential to pay special attention to the requirements associated with small-scale and medium batch production by designing the manufacturing process in both a rational and economic manner.

Without sacrificing flexibility, an improvement in production may be achieved by the application of automated components that are based on free-programmable control and data processing systems. Developments till today have mostly been restricted to special solutions for manufacturing problems, so that several independant key-points for automation have been determined within a single factory. The future task lies in the automation and further development of various production islands under consideration and their dependancy on each other (Fig. 1). Beside the machining process itself, it is therefore necessary to consider handling and transport functions as well as measurement and quality-control tasks involved in process automation. These play an important role in modern developments for the automation of complex machine set-ups that involve a physical and organizational interlinkage of several machines.

A manufacturing plant for parts production rarely is set up in one step in Europe. Certainly, it is not possible to automate the complete production of a factory in one step.

Depending on the manifold and possible extension stages all production - commencing from feeding of machines by handling devices and their interlinkage by automatic workpiece transport systems to flexible manufacturing systems -, may undergo a gradual and economic automation process.

The primary experience gained in the past from published installations, that are partly existant in a few places in industry have not been able to attract general interest. Some of the points that have affected general acceptance are given below:

- investment risk
- general long innovation time for complex systems
- limited flexibility with respect to user requirements
- insufficient user experience
- poor training of qualified production personell.

This situation has led to a project within the 3rd German Computer Development Program aiming for a modular approach with standardized ports in hardware and software in order to overcome some of the mentioned problem areas.

The problems encountered have three main reasons:

- problems in the mechanical processes
- problems associated with the hardware which usually is delivered from several suppliers
- problems by the handing over and generatio of the control and organizational software.

Note that these problems are discussed in detail by Weck, Zenner and Tüchelmann (1979)

In the production area one can distinguish between a Technical and an Organizational Data Flow. The following considerations deal mainly with problems for these data flows and of quality control a key point in the automation. The explanations and descriptions are based on an advanced Direct Numerical Control (DNC) System, this system is presently expanded to include operating data acquisition. In this overall production control system a multi-processor approach for automation of quality control will be integrated. The system is implemented at Pierburg Luftfahrtgeräte-Union, Neuß/Germany.

THE COMPANY AND PRODUCTION

The Pierburg Luftfahrtgeräte-Union with 385 employees is heavily engaged in the development, production and service of aircraft engine fuel control systems, precision flow meters for small and extremely small flows, pumps and automated test- and measurement systems. Productivity and flexibility in the production area is based on 14 NC-machines including 5 NC-machining centres with tool and palett changers.

The various problems of production data processing are solved by an IBM computer (370-125) with a modular structured programming system. The key point is a comprehensive data base which can be accessed by terminals from all main points in the plant. This data base consists of a number of distinguishable data sets, which can be handled by the available software system to give for instance a forecast of the production flow.

Access to Numerical Computer aided Part Programming Systems are available via modem connection from IBM-CALL and FIEDES-TELCOS in Switzerland.

The use of the commercial computer is limited as far as real time data handling is concerned. The implementation of a DNC-system has marked a major step of automation for the company with respect to the demands of the Technical Data Flow.

MODULAR DNC-SYSTEM

In a system for Direct Numerical Control (DNC) several NC machine tools are connected to a central computer via a data transmission system (Fig. 2). The primary task of the process computer is to distribute control data of single part programmes from a central mass storage device to the individual NC-Machine tools. In this way the administrative and distributive work involved with handling paper tapes of the NC-Machine tools can be avoided. The set-up of such control systems, eliminating the disadvantages of the tape reader (BTR-System, Behind Tape Reader) was the first step in the construction of computer controlled manufacturing systems.

Furthermore, in these systems other elements can be integrated such as acquisition of operating data, material flow and quality-control.

System Architecture

The user, who, for example, decides to build a DNC-system, has the difficulty to suitably link the various components such as machine tools, NC-units and manufacturing computer to each other after having selected them, i.e. he has to find a solution at the level of process peripherals. The requirements are twofold: safe, reliable and at the same time simple and fair-priced data transmission systems on the one hand as well as user-specific, on-the-spot process interfacing systems on the other. In view of the overall system performance it is essential to consider such general aspects as modularity, flexibility and standardization. Another aspect is to have a fine granularity of growth increments to adapt to the individual needs and allow for further expansions. The specific features and the efficiency of the today available international standardized CAMAC System (IEC 516) with its data transmission capabilities fulfill the requirement to a high degree. With CAMAC the user has the possibility of using compatible hardware from various manufacturers within a system with a minimum of mechanical and electrical problems. Pilot implementations of the system are described by Zwoll, Zenner, Möhl (1976) and Zwoll, Müller and Becks (1977).

The following characteristic features mark the DNC-system (see Fig. 3), which has been implemented at several plants:

- CAMAC module system as standard process peripherals for coupling numerical control of machine tools, connecting terminals, integrating auxillary functions - for e.g. acquisition of operation data and control of material-flow.

- CAMAC Serial Highway as data transmission system for bridging large distances between manufacturing computer and process.

- High efficient mini computer as manufacturing computer that processes PROZESS-FORTRAN using a real-time operating system.

The software costs of a DNC system often overwhelm those of the hardware components so that similar demands as those made for the hardware are also valid for the software. Therefore, the aim for standardization must be followed by using high-level problem oriented programming languages for the user software.

For the implemented system Process-FORTRAN has been selected. Fortran proves to be particularly advantageous for this application field, because it is wide spread, popular and easy to learn.

Hardware Components

The modular CAMAC system offers a good choice for realizing user like functional units. These can be used for implementing auxiliary functions like material flow control, acquisition of measuring and operating data as well as for coupling existent machines via DNC adapters. Paper tape readers used for supplying data to control units - NCs and CNCs - must be substituted in their function when connected to a manufacturing computer in a DNC-system. A special CAMAC-DNC adapter was developed for this purpose. By utilizing the advantages of modular techniques, it was possible to achieve a consequent separation of control functions (BTR Module) and communication functions (TTY Module) (see Fig. 4).

The BTR Module serves as an intermediate buffer for NC control data received from the manufacturing computer and also as a standardized interface (EIA-RS-4o8/9) for coupling NC control units.

Software System

The base of the DNC Software System is a real-time multiprogramming, multiuser operating system with extensions for the CAMAC-Driver and a data File Management (DFMS) system (see Fig. 5). An efficient Task Management yields to a highly structured modular user software system. The CAMAC Driver influences to a high degree the real time behavior of the whole system. An interrupt (LAM) handler is implemented which conforms with the operating system and a package of a few CAMAC-Subroutines using standard subroutine linkage to the high level Process-FORTRAN language makes operation easy. The user is not concerned with details of interrupt service routines and system conventions (for example LAM's are handled by setting eventflags). In this way programming of the CAMAC process peripherals is simple. Furthermore, the error recovery procedures of the CAMAC driver in conjunction with the installed bus coupler, serial crate controller and driver garanties fail safe operation within the Serial Highway. In combination with the host operating system the data file management system is an efficient and flexible facility for data storage, retrieval and modification under real-time conditions. Especially in a DNC system where a lot of application dependent files are to be handled in an efficient and economic way, this is one of the key-points. Thus files can be organized sequentially, relatively and by the indexing method, such systems are also known as Record Management Services (RMS-11 for DEC PDP11 Systems). In this way assembly-programming is avoided as far as the process peripherals and the organization of user dependent files are concerned.

All user tasks are written as mentioned above in Process-Fortran in a modular structure. Besides a set of basic functions, for the management of NC-part-programs and distribution of NC data, the DNC program system includes a series of extended functions, such as supervision of machining time, correction of NC data and communication functions. User-specific function modules can easily be added to this basic version.

ACQUISITION OF OPERATING DATA

In the past operating data acquisition systems have been implemented independent of the application of process computer in manufacturing. It has been influenced to a high degree by commercial data processing. This fact led to systems, which are very comfortable and user friendly, comparable with modern EDP Systems. Intensive software support, especially for interactive operating via a display terminal, transaction-manager with data error recovery and compilers for the commonly used higher level languages are available. But the flexibility is limited where an on-line connection to the machine tools or information exchange with the Technical Data Flow is required. A DNC-System can fulfill these requirements, also many application dependant modules are offered by several component manufacturers. The CRT-Terminal provided for DNC purposes can also be used for the acquisition of operation data. A link to the IBM computer is supplied by the minicomputer manufacturer. Thus a remote job entry (RJE) Terminal can be simulated by the process computer. The available minicomputer operating system with multiprogramming facilities, Data Management System and suitable compilers provide an efficient tool, to solve the problems of operating data acquisition. The central computer can be released from lower level data processing tasks (e.g interactive communication with operators in the plant, plausibility checks and data reduction). A further distribution of computing power can be achieved with the aid of a μ-computer installed in the process peripherals.

Thus a hierarchically structured global data processing system can be implemented with the typical features of real time functions combined with a comfortable data management. The CAMAC Serial Highway acts

as the central, high reliable data transmission link between the distributed computing power in the manufacturing area and the central process computer. Fig. 6 shows the overall plant control system realised with two Minicomputers, which are coupled together via a fast bus and two mass storage devices (discs). In this way a clear separation of time critical tasks can be achieved, which are exclusively handled in the process computer 1.
The 2nd minicomputer acts as a coupling device to the commercial IBM computer and in conjunction with the discs as the central on-line data base. It is also used for software development and for standby redundancy during routine maintenance of the process computer 1, thus allowing continous production.

A simple TTY Module was used in the beginning as communication interface. With the realisation of operating data functions this module will be replaced by a smart µ-processor based module. This module serves for data reduction and image processing. Thus the data transmission to the process computer can be handled blockwise, which essentially reduces the system load. For fail safe operation it is possible to buffer the input data for certain amount time.

The commercial IBM computer acts as the central data base for the whole production and gives a forecast of production data via the computer link. These data sets are actualized at the process computer level in respect to the real conditions (e.g. machine tool failure). From production area the actual values are given by the operators and transmitted back twice each day to the IBM Computer for updating the management information.

AUTOMATION OF QUALITY CONTROL

In the production of aircraft components rigid quality standards and test certificates are common practice. This demands an extensive data acquisition and long time recording of test data. The quality control is an considerable part of the overall production costs. This has led to considerations for automation of the quality control area.

Test Object

The fuel controller is an essential part of the control system of an advanced aircraft jet engine. Figure 7 shows a controller of this type under final test.

It is a hydromechanical device with the main functions of boosting the aircraft fuel with the aid of a high performance gear pump to the main burners. In addition it controls the starting and cut out processes of the turbine and monitors various process variable limit values.

Figure 8 schematically shows the construction of the main engine fuel system. On the left hand side one can see the inlet from the wing tanks and on the right the outlet to the main burners. The input variables are various air pressures (P3, P4), a mechanical speed signal (NH) and several electrical voltages determined by means of an electronic computer (e.g. lane 1,2) which in turn are functions of pressures, speeds and temperatures of the thermodynamic process. Output variables are fuel rates and injection pressures.

The main pump is driven by the shaft of the high pressure section of the gas turbine via a reduction gear (NH). It delivers a volume proportional to the speed which cannot be influenced by the regulation system. Flow regulation takes place by recirculation of the excess delivered at any time and its return to the fuel supply system. A typical function of the system is the control of turbine acceleration to a higher speed when the aircraft pilot requests increased power. By using the pilots throttle lever an electrical voltage is varied which is applied to the electronic control system. Taking into account various process variables whose inputs are not shown in the block diagram, the computer incorporated in the electronic control system forms an output signal which in turn is related to the acceleration control by an electrical current change.

The electronic control system including signal transmission is duplicated for safety reasons (thus two separate channels). The acceleration control generates via an electro-pneumatic and a pneumatic-hydraulic transducer an output pressure which positions the variable metering orifice. This double signal conversion is necessary to take into account the proportion of the two pressures P4,P3, which is characteristic of the instantaneous operating state of the compressor. The change of flow cross section in the variable metering orifice as a function of the positioning displacement is produced by means of a calibrated profile. The excess fuel to be passed to the main burners for acceleration of the engine is in addition limited by the instantaneous speed of the shaft of the high pressure section. In order to take this relationship into account the pressure drop above the set cross section of the orifice valve is kept proportional to the square of this speed with the aid of the pressure drop unit. Thus in accordance with the law of flow through an orifice plate the volume of fuel flowing through the predetermined cross section becomes proportional to the speed of the turbine.

Based on the simplified block diagram (Fig.8), Fig. 9 shows the measured quantities and control media of the fuel control system. These are essentially as mentioned above: control currents, speeds, air and fuel pressures. The input variables are adapted from outside with the aid of the signals described, the measured quantities required for test procedure being recorded on various modules. Several parameters are adjustable by means of mechanical calibration devices (see Table 1).

Test Function

The <u>final acceptance testing</u> of the units in the test bay is of considerable significance. The acceptance test procedure is used to proof the contractually assured properties of each finished unit. Final acceptance testing of the units is an essential part of quality assurance, includes additional test cycles as evidence of reliable and reproducable functions. By this means it is also the intention to pick up faults which can creep in at the various production stages, in spite of all extensive quality assurance measurements. Also included will be **tests with increased test loading**. Final **acceptance** testing is preceded by extensive <u>regulation procedures</u> on the above mentioned setting devices in order to obtain a **standard product** acceptable to the contractor.

During the course of their life time the units are generally returned to the manufacturer several times for repair and periodic major overhaul. During these procedures the same checks and acceptance tests have to be repeated whereby reference to the original documentation will provide important indications as to changes which may have occurred in the meantime. For this reason access to the test data on a <u>long term filing system</u> forms an essential part of the test functions.

Test Procedure

The adjustment procedures, outlined above, final acceptance testing and the filing data acquired ask for a complex test procedure. As a dynamic simulation of the operating states of the power unit would involve an economically unacceptable outlay, the test procedure is limited to a large number of static test points whose sequence is laid down in binding test instructions. The required input variables are generated with the aid of a test rig especially designed for this purpose. For basic calibration it is necessary to run through and regulate approximately 40 different functions and characteristics. Per function 3 to 12 test points are set, with different input variables and parameter combination. 15 different correcting and measured variables are monitored. Table 1 gives a list of the measured and correcting variables of the test rig and of the fuel control system.

In order to obtain an optimum setting in an iterative procedure the sub-functions must be run several times. All values should be recorded in a comprehensive report. The reports are filed and kept available for the entire service life of the unit. During a checking and acceptance test up to 300 parameter setting operations are carried out and approximately just as many measurements are read off and recorded.

After completion of adjustment operations follows the actual acceptance test procedure where all functions of the test object are run without interruption in the specified sequence. A number of selected subfunctions are then repeated for reasons of safety in the presence of a quality control inspector.

Test Rig

Figure 1o shows a conventional test rig such as was previously used for carrying out test functions and procedures. To simulate the turbine parameter factors one requires a series of additional devices. These include supply pressure, pumps, compressors and vacuum pumps as well as a DC motor to drive the high pressure gear pump integrated in the test object. In addition signal generators will simulate the parameters of the electronic control system. A system for stabilization of the temperature of the test medium will be used on which depends the accuracy of the flow measurement.

The test medium is aircraft fuel. The test bay is therefore an explosion hazard area. All test and measurement devices and their installation must comply with safety regulations.

Reasons for Automation

High speeds and flow velocities in the test system produce a noise level which necessitates constant wearing of ear protectors.

The stress imposed on the test personnel is extremely high for the stringent requirements due to reliability and responsibility. This is aggrevated by noise and other ambient conditions. On account of the high invested capital value of the test rig system multishift operation is indicated in the interest of economical plant utilization.

It is difficult to access the test reports prepared and filed by conventional methods for systematic evaluation from which important information could be obtained for development of the product, the production methods and of the quality assurance methods. The manual administration necessary for this purpose is hardly acceptable economically.

From the small amount of information given it is clear that for these types of test and quality assurance functions it will be urgently necessary to look for possibilities of automation. The automation project in mind should be introduced step by step and through the subsidiary targets listed below should lead to an improved solution of the problems outlined in the description of the test procedure.

a) Profit improvement by acceleration of test procedures, increased utilization of plant and reduction of manual involvement;

b) reduction of quality fluctuations by reducing manual intervention and possibilities of error when reading and transmitting measured values;

c) improvement of working conditions in the test area.

Requirements

The automated test system should be designed in such a way that its essential sections are capable of extension and can be utilized for comparable applications. Particularly high requirements are imposed on flexibility and thus on the modular construction of the system so that the test rigs can be converted for changes in the test procedures at reasonable expense and if possible with the company's own equipment.

The implemented system should permit largely decentralized computerized test procedures and when in operation, it should be incorporated in the existing production control system. Figure 11 again illustrates the general problem definition of the automation project.

In an initial step it is intended to automate the test rig itself and data acquisition of all measured variables to be computer assisted. Here it must be possible to initiate the complete acceptance test procedure, running through certain sub-functions and selection of a particular test point from a control console. The setting operations on the fuel controller will continue to be carried out by an operator.

In a second extension stage consideration is being given to include some of the adjustments to be carried out on the test object itself in the automatic test procedure with the aid of suitable actuators. However, we are working on the basis that the first step which provides for corrections to the test object being carried out by hand will economically justify the outlay necessary for its implementation as the sub-targets already mentioned can be achieved to a very high proportion.

On account of the requirement profile the functional units necessary for process control have been shown individually in Fig. 12.

As interfaces to the existing system it will be necessary to install suitable actuators and adaptors on the test rig side. On the test rig itself there should be an operating panel while the other components should be accomodated outside the noise area within the explosion proof zone. A video display will be integrated in the operating panel in order to have a quasi analogue representation by means of measured value bars. This will replace the large number of analogue instruments used at present. In addition it should provide operator guidance and manual intervention in a test program in progress.

It is the intention that the control system be linked to the process computer via the transmission system (CAMAC Serial Highway). The "control supervision" unit with its input/outputs takes the form of a bit handling processor which, similar to a programmable logic control system (PLC) monitors the Boolean conditions necessary for the test procedure in question.

For adjustment and stabilization of various parameters (speeds, temperatures, air pressures etc.) a total of 8 digital controllers will be necessary; for the second stage of automation mentioned above additional stepping motor drive systems are necessary.

Through an analogue input approximately 10 measured values will be available for logging in addition to the controller variables available.

In the "test schedules and data processing" functional unit the individual test sequences are run consecutively and the data is logged and simultaneously represented in graphical form on an appropriate graphic terminal. The master console is also used, in the case of service and overhaul work on a fuel controller for representing test data obtained earlier, which is requested from the data file. A mass storage unit allows operating of the test bay even without the main computer.

Figure 12 shows the control system structured in modular form using complete functional units. Basically these can be realised in the form of dedicated individual boards in a PC-board system with a suitable bus. For reasons of flexibility, of possible re-utilization - also in control systems of other types - and thus of maximum economy of use it is adviseable to limit the problem-specific hardware outlay to indispensible matching functions. The application orientated functions can then be implemented software-wise with the aid of one or more micro-computers. In this way it is easier to achieve mass production of standard card types.

Realisation of the Test Rig Control

Microcomputer boards (OEM computers) are available off the shelf in a large number of types, offered by several manufacturers. This varity of modules allows to plug together a microcomputer system taylored to the needs.

Modularity is also advantageous in cases of concept changes (only a small part of the components have to be exchanged) and for investment splitting by stepwise system expansions.

Distributed computer power increases the overall system reliability. A failure in one part does not necessarily block the total systems. Defective modules can easily be replaced. During break downs operation can be continued partially.

A possible solution of a multi-microcomputer based system for the overall test rig control is shown in Fig. 13. It would fulfil the technical demands, but in an uneconomic way, due to the low load of the part of the system dedicated to test scheduling and test data recording. The master console, together with the test schedule computer, the diskette controller and the communication controller are mainly active in the beginning and at the end of a test run-off. In the meantime - which is about 9o % of the operating time - they are dormant and would only be activated in emergency cases.

In contrast to this, the part of the system providing the functions of data acquisition and remote control are heavily loaded during most of the operating time. Especially the activities of the eight digital control loops invoce a heavy bus load.

The unequal distribution of system activities urges to split the system into two separate parts as shown in Fig. 14. The upper part includes all those boards necessary to program the test schedules, to transmit these to the test rig remote controls and to record and process the test data. As the system load of this unit is comparatively low, it can be used to provide these functions for several test rigs. The test rig remote control units -three in this example - are equipped for the tasks of data acquisition, control of the test parameters (8 loops) and the operation control panel. A serial data link connects the remote control unit to the central control unit.

Figures 15 and 16 show more details. The heart of the data acquisition and remote control unit is the operation control μ-computer. It arbitrates the operation panel, i.e. displays measured values and operational information on a CRT and accepts commands from the operator. Further it supervises the total system activities such as the operator dialogue and the generation of the communication-protocol with the central control unit via the communication controller.

The intelligent analog I/O interface collects testdata and input values for the parameter control algorithms. The 4-2o mA signals are converted to 12 bit hex numbers and put in a mailbox RAM for the use by the μ-DDC computer or the operation control computer. The μ-DDC-computer calculates the manipulated values by PID-algorithms which drive the analog output board. The digital I/O monitors binary inputs such as switches, emergency stop buttons and outputs control current to relays. The memory board is used for program and data storage and as a mailbox for inter-computer-communication. All boards are connected to a multiprocessor bus.

The configuration of the central control unit is shown in Fig. 16. The system is governed by the test schedule computer. On this board not only the test schedules are programmed, but also the test data are processed for displaying on a graphic terminal and recording on floppys as a temporary mass storage medium. For long-time storage the data are transmitted serially via the Serial Highway interface, based on a fast bipolar SMS 3oo μ-Processor to the central IBM 37o-125 computer. The communication controller in the lower part of the figure provides the serial data interchange with the remote control units of the test rigs.

COMPARISON OF MULTIMICROPROCESSOR SYSTEMS

Three Multi Microcomputer Systems, which would be suitable for the implementation of the control unit, were considered in detail:
    INTEL iSBC-Multibus
    AMS-Siemens
    MPST

The well known INTEL iSBC-Serie is originally based on the 8o8o CPU. For more details see INTEL Publications. The system has become very popular and is now supported by many other manufacturers.

Siemens, second source for many INTEL Products in Germany, introduced the AMS-Bus, which was derived from the Multibus with several useful, future oriented facilities added. Extensive technical papers have been published by the company SIEMENS. Bieck (1979) discussed the production program of AMS-boards and the future plans of expansion.

The iSBC-Multibus and AMS-Bus Products have both been designed specifically to work with INTEL/Siemens Microprocessors. Therefore bus configurations and timings functionally support these processors.

The MPST-Bus had been developed by a group of German machine tool manufacturers in collaboration with research institutes and universities. Stute (1978) and Wörn (1978) gave a comprehensive overview of the system. It is taylored for the specific needs of machine tool control. The work is supported by the Government within the 3rd German Computer Development Program (PDV Projekt). One of the basic ideas of the MPST was, not to limit it to one type of microprocessors but to design it processor-independant.

The following chapter will give an abstract of the differences between the three Bus Systems and their suitability for machine control including DDC and data acquisition.

Comparison of the Bus Designs. As the Multibus and AMS-Bus are very similar, only the main differences between these are pointed out:

The AMS-Bus functions can be considered as a superset of the Multibus. The AMS-Bus is expandable upto 32 data lines and 23 address-lines. It consists of a "System Bus", which contain all lines necessary to operate a Multiprocessor-System with 8 bit data and 16 bit addresses. The "Extension Bus" includes those lines needed for 16- or 32 bit data transmission and an address space of 8 MBytes.
Both bus systems INTEL and AMS provide means for effective ROM-RAM or Memory-I/O override, by the aid of separate memory and I/O-command lines, which allow alternative use of parts of the addressspace.

Interrupt recognition can be handled by bus vectored or non bus vectored interrupts with the 8 interrupt lines. The AMS-Extension Bus additionally allows to implement two user defined interrupt lines. Interrupt priority is either resolved in parallel or in serial (daisychain) on user's choice. Data error detection is facilitated by two parity lines on the AMS-Bus (for higher and lower byte). With these facilities the Multibus and the AMS-Bus can be used for numerous present and future micro-computer applications.

The MPST-Bus, which has been dedicated to machine tool control, as outlined (1979) by Stute, Spieth and Wörn, does not provide such extensive facilities. Its data word length of 16 bit is completely satifactory for this purpose. The addressspace, originally limited to 16 address lines, can be expanded by an overriding technique, which allows to address a maximum space of n x 64 kByte, where n is the number of boards supplied by the bus.

Interrupts are distinguished by their bus vectors. Priority arbitration is done by daisy chaining.

An important difference between the Multibus/AMS Bus on the one hand and MPST on the other is the method of bus arbitration.

On INTEL/Siemens-Products a bus arbitration logic is incorporated on each master-board (that is a board with the ability to control the bus, in contrast to a passive slave). A master which wants to use the bus, monitors the Bus Busy line. If it is inactive and no other master with higher priority claims the bus, Bus Busy is set and the interaction is started. After its completion the master releases the bus again.

The MPST-System uses a dedicated bus manager on a separate board to coordinate bus operations. A master claiming bus access signals this to the manager by an interrupt. If no higher priority interrupt occurs, the manager will grant the access. The active master then sets the Bus Busy flag, which is monitored by the manager. It is reset after the action is completed.

Mechanical Specifications The INTEL Multibus provides two rows of connectors, a main connector with 86 pins and an auxiliary connector with 6o pins. Boards interfacing the Multibus have standard outlines of 12 x 6.25 inches.

The AMS-Bus and the MPST-Bus specify boards with the European Standard outlines called "Double Euro Card" of 233.4 x 16o mm^2 (= 9.19 x 6.3o inches). AMS uses two connectors with 96 pins each.
The MPST-specification allows two connectors containing 32 pins each. Upward compatible 96-pin connectors may be applied by user's choice, but specifications for this are not given.

Software Aspects. In the application, three one-board-microcomputers will operate on a common system bus. Unfortunately, an operating system which would powerfully supply multicomputer system functions, is not yet available from the µ-processor manufacturers. SIEMENS announced an advanced Multitasking Multiprocessor Real Time Operating System which will be available in spring 198o. The MPST developers intend to provide fixed software interfaces for machine tool control applications. However, studies on this item are still in the beginning phase.

State of Implementation. The INTEL-Multibus System has become a defacto standard because of its world wide use and support by manufacturers of various expansion boards. The AMS-Systems has recently been introduced to the market, the program offered is not yet as comprehensive. Therefore the user himself has to build several functions on an universal interface board.

Several prototype versions of the MPST System actually work at various institutes participating in its developments. The MPST-Computers serve a variety of peripherial I/O-boards specially designed for process- and manufacturing control. Efforts are made to standardize the MPST-Bus in Germany.

Some machine tool manufacturers have developed there own µ- computerized numerical control around the MPST-Bus in Germany.

Selected Microcomputer-System. For the task of the test rig the AMS-Siemens System has been chosen. The reasons for this decision are as follows:

1. The AMS-program includes a One-board-microcomputer with an on-board arithmetic

processor, which is very useful for comfortable and fast calculation of digital control algorithms. In the application fast mathematic operations are a substantial need and extremely timecritical. Therefore the on-board math unit offers a considerable advantage, as the interaction between the CPU and this unit does not affect the system bus and thus can not be held up by other systems activities.

2. The AMS-System is the only Multi-Microcomputer system presently available off the shelf, that uses the Euro-Card standard, which is applied by most European manufacturers and users. Thus on the one hand a lot of supporting hardware (cages, racks etc.) are available and on the other hand financally powerful users (like the Federal German Railroad) demand products of this standard. The AMS-System therefore has a good chance to gain a high popularity in Germany.

3. As the AMS-Program is still in the phase of introduction to the market, the chance is given to install it in a pilot project. The experiences out of this implementation (the development of user-specified boards and a multiprocessor operating system) can be used in future projects.

CONCLUSION

Figure 17 shows the overall production control system at Pierburg Luftfahrtgeräteunion. At present 10 numerically controlled machine tools are linked to the process computer. Operating data acquisition is being implemented. Based on the operating experience with the DNC-System, data acquisition of the heat treatment process will be included.

The use of the modular standardized CAMAC system has been very attractive for the company. It allowed a stepwise expansion of the basic DNC-System, thus minimizing the financial risk. The companies Softwaregroup is pleased with the implementation using Real Time Fortran. This gave them an easy opportunity to modify the software as production requirements changed.

The cooperation of Institutes, Computer Manufacturer and a potential user during the implementation was very helpfull to structure the system in a way that it can also be used for other installations.

It is expected that the Multi Microcomputer approach for the automation of the quality control area, will result in a solution which is applicable and easily adaptable for other installations.

REFERENCES

Adams, G. and Rolander, T. (1978). Design Motivations for Multiple Processor Microcomputer Systems. Computer Design, March 1978.

Bieck, R. D. (1979). AMS, das erste Baugruppensystem aus Europa für Multicomputer-Anwendungen. Siemens Bauteile Report 2/79. Munich, Germany.

Electronic Industries Association. (1979). EIA Standard RS-408, Interface between Numerical Control Equipment and Data Terminal Equipment. EIA-RS-408, Electronic Industries Association, Engineering Department, 2001 Eye Street, N.W., Washington, D.C. 20006.

INTEL (1975). Multibus Specification. Pub. 98 - 638, INTEL Corporation, Santa Clara, Cal., USA.

SIEMENS (1979). The AMS-Bus, Siemens Technical Publication. Munich, Germany.

Stute, G. (1978). Grundgedanken von MPST. Informatingstagung MPST am 23.2.78. Hrsg. vom ISW-Institut für Steuerungstechnik, Stuttgart, Germany.

Stute, G., Spieth, U. and Wörn, H. (1979). A Multiprocessor Control System as an Universal Modular System for the Design of Machine Tool Controllers. Pergamont Press Headington Hill Hall, Oxford,U.K.

Weck, M., Zenner, K. and Tüchelmann, Y. (1979). New Developments of Data Processing in Computer-Controlled Manufacturing Systems (DNC, FMS). CASA-SME, Technical Paper MS79-161, Dearborn, Michigan, USA.

Wörn, H. (1978). Darstellung des MPST-Bus, Tagungsunterlagen MPST - Modulares Mehrprozessorsystem am 23.2.1978. Hrsg. vom ISW-Institut für Steuerungstechnik, Stuttgart, Germany.

Zwoll, K., Müller, K. D., Becks, B. Erven, W. Sauer, M. (1977). Direct Numerical Control of Machine Tools in a Nuclear Research Centre by the CAMAC-System. IEEE Transactions on Nuclear Science, Volume NS-24, 1, S. 446.

Zwoll, K., Zenner, K., Möhl, R. (1976). Application of Standardized Process-Peripheral System (CAMAC) for Direct Numerical Control of Machine Tools, Proceedings of the Second International Symposium on CAMAC in Computer Applications. EUR 5485 d-e-f, p. IV 1-3, Brussels, Belgium.

(Editor's Note: Due to an electronic failure in our transcription equipment, the discussion following the presentation of this paper was lost and could not be reconstructed. We apologize to the authors, participants, and readers.)

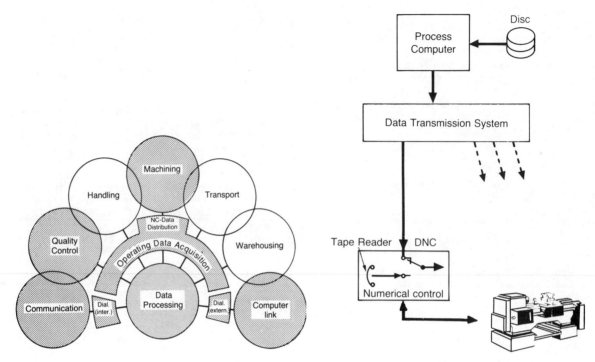

Fig. 1 Application Fields and Functions of Process Data Control in Manufacturing

Fig. 2 <u>D</u>irect <u>N</u>umerical <u>C</u>ontrol (DNC)

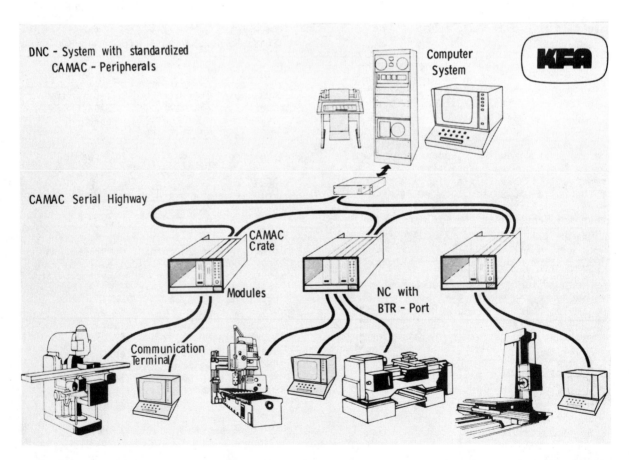

Fig. 3 DNC-System with standardized CAMAC-Peripheral

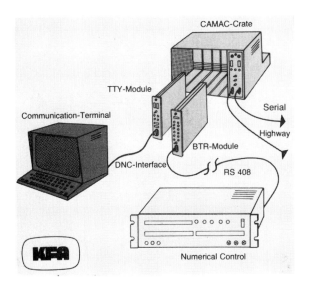

Fig. 4   DNC-Interface by the CAMAC-System

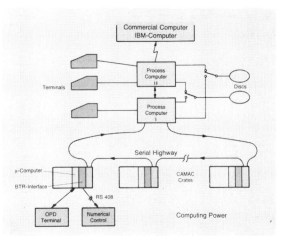

Fig. 6   Structure of the Process Control

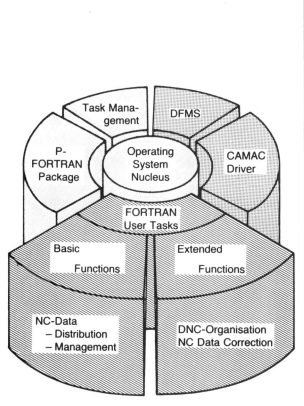

Fig. 5   DNC-Program System

Fig. 7   Fuel Controller under Final Test

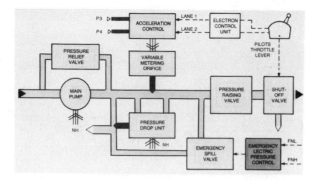

Fig. 8   Main Engine Fuel System

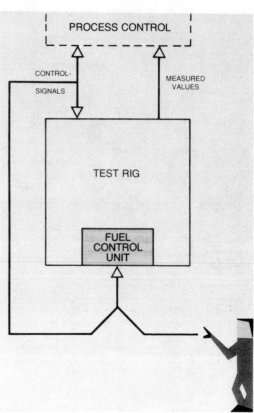

Fig. 11   Test System Schematic

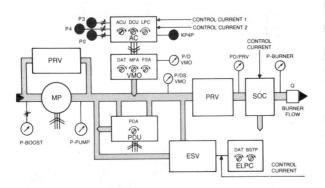

Fig. 9   Measured Quantities and Control Media

Fig. 10   Conventional Test Rig

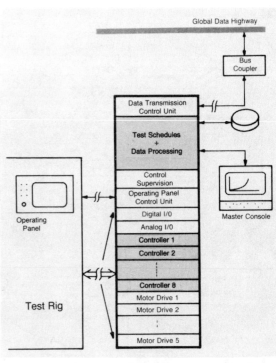

Fig. 12   Functional Units of Process Control

Automation of Quality Control 117

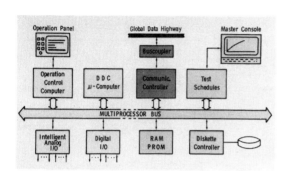

Fig. 13 Overall Test Rig Control with a Common Multiprocessor Bus

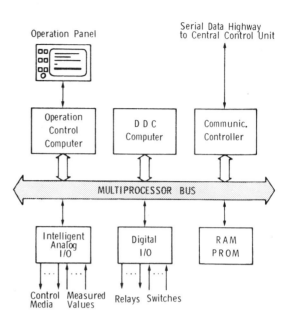

Fig. 15 Data Acquisition and Remote Control Unit

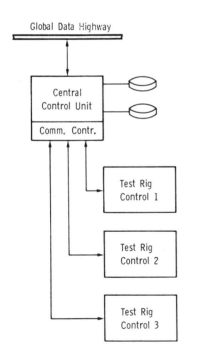

Fig. 14 Test Rig Automation System (General View)

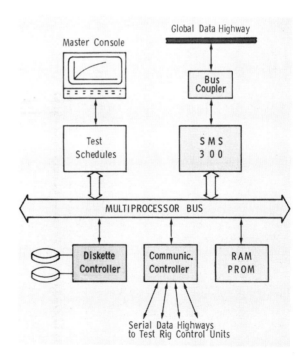

Fig. 16 Central Control Unit

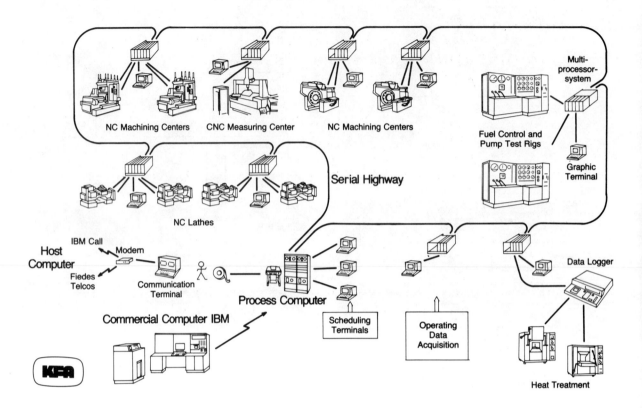

Fig. 17 Overall Production Control System at Pierburg Luftfahrtgeräte Union GmbH

Table 1: Measured and correcting variables of text rig and fuel controller

	Control media		Measured quantities	
Test rig	1. Speed 2. Air pressure 3. Air pressure 4. Air pressure 5. Fuel pressure 6. Control current 7. Control valve 8. Control valve 9. Selector switch 10. Selector switch	NH P4 P3 PØ PBOOST IXLPC VX VY Channel 1/2 MLPC/ELPC	1. Speed 2. Air pressure 3. Air pressure 4. Air pressure 5. Fuel pressure 6. Control current	NH P4 P3 PØ PBOOST IXPLC
Fuel-Control-System	1. PDU adjuster 2. VMO Datum adjuster 3. VMO Max. Flow adjuster 4. VMO front stage adjuster 5. ACU adjuster 6. DCU adjuster 7. LPC datum adjuster 8. ELPC datum adjuster 9. ALPC backstop		1. Fuel pressure 2. Fuel pressure 3. Fuel pressure 4. Fuel pressure 5. Fuel pressure 6. Fuel pressure 7. Burner flow	PBOOST PPUMP PBURNER B/DSVMO PDVMO PD/PRV

CONCEPT AND EXAMPLES OF A DISTRIBUTED COMPUTER SYSTEM WITH A HIGH-SPEED RING DATA BUS

T. Kawabata* and H. Tanaka**

Software Design, Industrial Computer Department, Toshiba Corp., Tokyo, Japan
**Industrial Computer Department, Toshiba Corp., Tokyo, Japan*

Abstract. To make effective use of the transmission speed and performance characteristics of our ring data bus, the concept of three networks has been devised. The three are a PIO access network, a BIO access network, and a message transmission network. The first two use the centralized processing method, and the last uses the distributed processing method. The distributed processors are further ranked according to their capabilities as what we call "host nodes", "cluster control nodes", and "sub-nodes".

The network functions are classified into four layers, and each has its own protocol. This article describes this concept, and a ring data bus based upon it.

Keywords. Ring Data Bus; Network; Node; Protocol; Process Computer

INTRODUCTION

Early in the process industry, computerized systems for the control of steel production, sewage disposal, etc. were constructed simply to automate basic processes or production lines. This was due to the fact that, at that time, the abilities of computers, including their responsiveness and expansionability were severely limited. With the rapid improvement in computer capabilities, small-scale computerized processing systems came to be combined into total systems. The objective was to use the central processing method to improve process control and management efficiency. However, system expansion has become quite difficult, since most large-scale total systems have higher loads than can be handled by their centers. In this situation, load or function distributed processing becomes more valuable than centralized processing.

In the meantime, the remarkable developments in LSI technology have drastically reduced the costs of high-speed semiconductor storage devices such as RAM's and ROM's. These reductions in cost have also resulted in the appearance of a variety of processors of high speed and reliability. Systems based on function distributed processing are increasing more than ever. This type of processing is regarded as improving the performance and reliability of the systems concerned, thereby adding new capabilities to them. However, these systems have a drawback; system management and maintenance is not an easy task. On the basis of our experience in the development and implementation of ring data bus systems, we have classified all the functions required in a new distributed processing system into two groups, those which must be distributed and those which must be centralized. We have then set up a model system in which various conventional processors and terminals, and also various access protocols, are logically consolidated. We feel that with this modal system: (1) the above-mentioned drowbacks can be removed, (2) neither the addition of a device to the system nor the change in a process will require any serious alteration in the system, and (3) resource sharing (the main objective of any network) will be facilitated.

EXISTING RING DATA BUS SYSTEMS

Thus far, two types of ring data bus systems have been implemented. In the first system (TOSWAY-1000), the process I/O devices at the distributed stations are controlled at the center of the system. Transmissions are performed from the center to the devices or vice versa (i.e. 1-to-N transmission). The main objective of this system is to reduce cabling costs. In the second system (TOSWAY-1500), distributed processors are connected to a transmission line or bus in the form of a ring. Information is transmitted at high speed from one or more processors to the others simultaneously (i.e. N-to-N transmissions).

First Ring Data Bus System

Figure 1 shows the structure of the first ring data bus system. The master station

(MST) synchronizes line control signals, controls data transmissions over the line, supervises the status of each station (STN), etc. Normally, the line, or a physical data transmission path, is not circular. If a station abnormality occurs, the line switch is operated to form a circular line, allowing data transmission to be performed in the reverse direction. Process I/O devices may be connected to each STN.
If connected, they are accessed by the single MST processor in either a cyclic access mode or an urgent access mode. In the cyclic access mode, data is input or output at fixed periods of time; in the urgent access mode, it is input or output immediately.

Second Ring Data Bus System

Figure 2 shows the structure of the second ring data bus system. The master station (MST) at which the host processor is located manages the entire system by performing (1) synchronization of line control signals, (2) control of data transmissions over the line, (3) supervision of the status of each station (STN), (4) detection of station failures and their isolation, (5) detection of cable disconnections and isolation of the disconnected parts, and (6) isolation or restoration of any particular station. The station at which a processor and a terminal controller are located may also control data transmissions.

The basic software system (DTW) at each station sends data to or receives it from other stations. In addition, the DTW at the master station can initialize the processor or terminal controller of a station when necessary.

Three different service software systems are provided with the second ring data bus system: DXW-I, DXW-II, and HTSS. DXW-I supports the transmission of a main memory image from or to a remote processor. DXW-II supports message transmission from or to a remote processor. HTSS controls the terminal controller at a remote station.

RAS conversational utility routines perform the following functions during interactions with an operator: (1) line activation, (2) line restoration, (3) station isolation, (4) station status detection, (5) setting up a link between the main line and the stand-by line, (6) down-line loading, and (7) line supervision.

CONCEPT OF A NEW RING DATA BUS SYSTEM

After reviewing the functions of the preceding two ring data bus systems, we determined that a new ring data bus system should use processors based on a 32-bit architecture are connected, that the ring data bus needs to be as flexible as possible so that processors based on different architectures (e.g., a 16-bit or 24-bit architecture) may be used, and that all processors so far developed can be used with it. Keeping these requirements and the following idea in mind, we embarked on the formation of a conceptual model of the new system:

We believe that when designing a network consisting of processors and terminals connected by lines, all the characteristics of the hardware and software (e.g. an operating system, programs written in various languages, user process interfaces, etc.) to be used must be considered. So, we have decided to systematically define the logical network in the basic concept as being free from these characteristics. It then becomes possible, we feel, to regard various processors, terminals, and user processes (or application programs), as logical components in a uniform manner.

Logical Network

The logical network is divided into the following three virtual networks (VN):

(1) Process input-output device (PIO) access network;
In this network, PIO devices can be accessed and shared by any processor in the same manner, whether the network is used or not.

(2) Business input-output device (BIO) access network;
In this network, BIO devices (e.g. I/O typewriter, cathode ray tube visual display unit, etc.) can be accessed and shared by any processor in the same manner, whether the network is used or not.

(3) Message transmission access network;
In this network, all resources may be equally shared by all processors. Also, communications are conducted between processors, or processes, and such communications make distributed processing possible.

The first two virtual networks are used for centralized processing, whereas the last is used for distributed processing. The three networks are illustrated in Figure 3.

Logical Nodes

To make clear what the structure of the logical network is logical nodes in the network are grouped into one of three types:

(1) Host nodes (HN)

Host nodes are processors which can independently establish one or more TP or FC paths (described below). A host node executes various application processes, e.g. control an industrial process, input data from or output data to a terminal, exchange data with another application process, etc. Among host nodes, some may supervise a virtual network, and others may perform complementary functions, or boundary functions (BF),

Distributed Computer System

enabling sub-nodes to operate in the logical network normally. In addition, a host node may access a PIO or BIO access network to use an I/O device in it. Where two different processes exist on a host node and a cluster control node respectively, the host node supports the inter-process communication.

(2) Cluster control nodes (CCN)

Cluster control nodes are processors which can control I/O devices in multiple information transmissions, execute the various application processes stated above, and establish one or more TP or FC paths. Every node of this type is logically connected to a host node.

(3) Sub-nodes (SN)

Sub-nodes are equivalent to processors in control functions. Any node of this type is logically connected to a host node and can have TP paths only.
A logical network illustrating the above is given in figure 4.

Addressing

All network addresses (NA) used in the logical network consist of two types of logical addresses, node addresses (NDA) and unit addresses (UA). A network address is in the following format:

| NDA | UA | Network address

The NDA consists of three numeric values: (1) the physical number of one of the stations in the logical network used for the maintenance of hardware there, (2) the number of a device control unit used at one of the stations, and (3) the number indicating the endpoint of the physical address of a hardware unit. A node address is assigned to each node for unique identification, regardless of type. It is also used to establish a TP path.

Unit addresses are used to uniquely identify three types of communicating entities, network logical units (NWLU), network management units (NWMU), and node management units (NDMU). A host or cluster control node, contains one NDMU and one or more NWLU's. The NWLU's in one node communicate with the NWLU's in another node, whereas the NDMU in a node manages the node itself. A host node may also contain one NWMU, which manages the network itself. A sub-node may contain one pseudo -NWLU which is assigned to a single TP path.

Combinations of these two types of logical addresses, or network addresses, make the unique addressing of nodes and their units possible in the logical network. Figure 5 shows NDA's, given to various nodes, and units, existing in the nodes.

Logical Structures Of Virtual Networks

PIO access network (VN1)

The PIO access network is a virtual network where the control of PIO devices (originally performed by the host node at a station) is assigned to another station connected to the ring data bus (i.e., a distributed function). This network may be accessed and used by host nodes only, and provides the same method of controlling PIO devices as if they were locally connected to the host node. This virtual network performs the function of a data link control layer only (see Figure 6).

BIO access network (VN2)

The BIO access network has the same features as the PIO access network, except that BIO devices are used in place of PIO devices.

Message transmission network (VN3)

A message transmission network allows the resources distributed over the three types of nodes to be shared efficiently and smoothly. This is the main objective of this network, and to achieve it, the communication functions have been logically defined to be divided into four layers (described later), and corresponding protocols have been determined. Figure 6 shows the four types of layers in a structured form. It should be noted that the host node and cluster control node have logically identical layer structures. Whether a node is a host node or a cluster control node is determined by the particular functions selected for it. Usually, the cluster control node has a function control layer capable of performing fewer functions than that of the host node. A sub-node has the data link control layer (DLCL) and the transport control layer (TPCL) only. This node is always logically connected to a host node in communications.

Layers

Data link control layer (DLCL)

This layer establishes a logical connection between the stations indicated by the NDA's, controls logical communication devices in the stations, and communicates with other DLCL's. A communication path established by this layer is called a data link (DL) path, and is provided as a transport layer. The DL path is divided into the following three sub-layers:

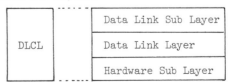

The hardware sub-layer is the part of the hardware controlling the ring data bus and acting as the interface with the line. This sub-layer is known as the line coupler.

The data link layer establishes logical connections between stations and serves to control communications smoothly.

The data link sub-layer supervises and manages the ring data bus. If an abnormality occurs in some part of the ring data bus, this sub-layer detects it and controls the line in such a way that the operative stations connected to the bus remain largest in number. This is called line restoration, and it is one of the most important functions of the ring data bus.

Transport control layer (TPCL)

This layer establishes a logical connection between the logical nodes indicated by the NDA's, and provides the means to transmit transparent data. It also, makes certain that a transmission from a source node to a destination node has been performed successfully. This is called the 'end-to-end significant' function. Data flow control is another function performed by this layer to make certain that information is transmitted from one logical node to another. A logical communication path established by this layer is called a 'transport (TP) path', and is provided as a function control layer.

Function control layer (FCCL)

This layer establishes a logical connection between the processes indicated by the UA's, and provides a flexible, logical communication path, called the 'function control (FC) path', to be used by an application layer. This is the path from one end process to another and is also called a 'session'. Additionally, this layer performs data flow control, such as session set-up and release, data chaining, response control, etc. A TP path may contain a number of FC paths (i.e. multiplexing is possible). All functions performed by the application layer are controlled by this layer with appropriate methods.

Application layer (APLL)

This layer is constructed on the basis of the functions performed by the other three types of layers, and performs various application functions when required.

Protocols

As shown in Figure 7, the layers in the logical network structure require their own protocols for the performance of their functions. The full protocal names and their abbreviations are the following:

    Data link protocol (DLP)
    Transport protocol (TPP)
    Function control protocol (FCP)
    Application protocol (APP)

Note that the APP is not defined in this conceptual model.

Data link protocol (DLP)

This protocol is used between DLCL's in communication with each other within a PIO access, BIO access, or message transmission network. A unit of data contained in this protocol is called a 'data link unit (DLU)', and is in the following format:

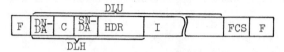

where

F: A flag indicating the beginning and end of a DLU.
DLH: A data link header consisting of:
DNDA: A destination node address;
C: A command used to indicate the attribute of the DLU;
SNDA: A source node address;
HDR: A header indicating control information defined seperately in the PIO access, BIO access or message transmission network.
I: A piece of information designating a TPU (described later).
FCS: A frame check sequence used to check the DLU.

In addition to the DLU format, this protocol includes <u>two main procedures</u> required for data link level communications. The first, needed to aquire the right to send, is illustrated below:

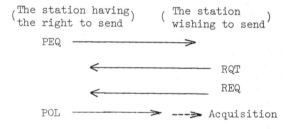

where

PEQ: A command used to inquire if a station wishes to send.
RQT: A command used to inform the inquiring station that the particular station does wish to send.
REQ: A command used to acquire the right to send.
POL: A command used to pass the right to send to the requesting station.

The second, needed to transmit data, is illustrated below:

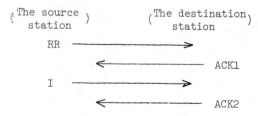

where

RR: A request asking the destination

station to receive the information.
ACK1: An affirmative acknowledgement to the request (i.e., the destination station is ready to receive the information).
I: The information transmitted.
ACK2: An affirmative acknowledgement to the source station (i.e., the destination station has received the information).

Transport protocol (TPP)

This protocol is used between TPCL's in communicating with each other. A unit of data used in this protocol is called a 'transport unit (TPU)', and is in the following format:

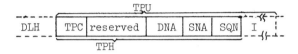

where

TPH: A transport header consisting of:
 TPC: A transport category indicating the attribute of the TPU;
 DNA: A destination network address;
 SNA: A source network address;
 SQN: A sequence number, attached to the TPU for sequential transmissions and used in the end-to-end significant function.
I: A piece of information designating an FCU (described later).

In practice, the TPH may be one of three types: TPH1, TPH2, or TPH3, depending on the types of nodes involved --- that is, TPH1 is used in communication between host nodes; TPH2, between host node and cluster control node; TPH3, between host node and sub-node. This is because each type of node has different characteristics in function.

Full-duplex transmission may be conducted using this protocol. The source node attaches an SQN to a TPU, then transmits it to the destination node. The destination node responds by transmitting a response TPU with an identical SQN to the sending node. Then, both nodes perform an end-to-end significant function to make certain that the transmission is complete. This protocol also places a limit on the number of TPU's capable of being transmitted at one time, and allows control of information flows to be performed by checking SQN's.

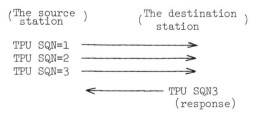

Function control protocol (FCP)

This protocol is used between FCCL's. A unit of data used in this protocol is called a ' function control unit (FCU)', and is in the following format:

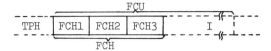

where

FCH: A function control header consisting of:

FCH1: Function control header 1 indicating the attribute of the FCU to be used either for a session or by an application layer;
FCH2: Function control header 2 indicating the attribute of the response method to be used;
FCH3: Function control header 3 indicating the direction of the communication.

This protocol also includes three procedures required for function control level communications: a session set-up/release procedure, a data transmission procedure, and a network management procedure.

The session set-up/release procedure is used to set up a session for communication between NWLU's, and release the session at the end of the communication.

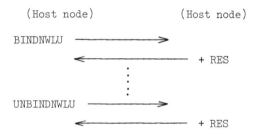

where

BINDNWLU: A command used to set up a session.
UNBINDNWLU: A command used to release the session.
+ RES: An affirmative response.

The data transmission procedure uses the following transmission control functions so that an application unit (APU) (a data unit to be used by the application layer) can be transmitted successfully: (1) data chaining, (2) control of appropriate response (i.e., either affirmative or negative), (3) control of appropriate transmission mode (i.e., either full-duplex or half-duplex transmission mode, depending on the right-to-send).

The network management procedure is used between an NWMU and an NDMU to check and select, or simply trace, a data unit transmitted from the one node to the other. In addition, this procedure is used for the purpose of loading processing programs from host nodes to cluster control nodes, or sub-nodes, in down-line modes, or for the purpose of dumping the memories of cluster control nodes or sub-nodes, on to host nodes in up-line modes. This procedure also includes testing of cluster control nodes, or sub-nodes, by host nodes.

Figure 8 shows the relationship between the layers, protocols, and data units, thus far described. Figure 9 shows the three types of data paths of the new ring data bus system.

FUNCTIONS OF NEW RING DATA BUS

The new ring data bus, being developed on the basis of the conceptual model already described, is presented in the subsequent discussion.

Hardware Functions

One of the most important objectives of the new ring data bus (TOSWAY-10M) is to allow information exchanges between a number of transmission devices utilizing the physical transmission path on a time sharing basis.

(1) Three physical transmission paths or "channels" are logically provided: a message transmission channel used by the BIO access network or the message transmission network; a PIO channel used by the PIO access network; a modem channel, not directly related to a particular type of network but providing a physical transmission path.

(2) The physical transmission path consists of a main ring data bus and a stand-by ring data bus so that physical transmission path abnormalities can be handled with flexibility.

(3) The physical transmission paths and the stations are always supervised by a single supervisory station (SVC). This station has the functions of a data link sub-layer in the logical network structure. It isolates hardware units or stations in failure from the rest of the physical transmission path, and can also isolate a particular part of the physical transmission path when a new station is added to the physical path. In addition, the load on the physical path is monitored by the station from time to time.

(4) The number of channels in use, as well as the speed of transmission over a channel may be changed. This makes the physical transmission path of any system optimal, regardless of the use of the system.

(5) All hardware units have self-diagnostic capabilities so that any abnormal condition in the units can be readily detected.

Software Structure And Characteristics

The system software is used to establish networks of three types (i.e., PIO access, BIO access, and message transmission) on the basis of the new ring data bus. This software is illustrated in Figure 10.

Software structure

The system software is composed of five programs: (1) a PIO executor program (PIOX), (2) an I/O control service program (IOCS), (3) a multi-purpose communication access method program (MCAM), (4) a MCAM-RAS program (MCAM-RAS), (5) a TOSWAY-RAS program (TOSWAY-RAS).

(1) PIOX is used to control PIO devices with the aid of the PIO access network. Any PIO device directly connected to a host node can also be controlled using this program.

(2) IOCS is used to control BIO devices with the aid of the BIO access network. Any BIO device directly connected to a host node can also be controlled using it.

Note that PIO devices and BIO devices are exclusively controlled by stations as resources in the PIO and BIO access networks. The networks themselves are supervised by TOSWAY-RAS (described later).

(3) MCAM is the general term for a program used to make a message transmission network accessible. This program consists of computer communication service (CCS) modules, which perform the same functions as a TPCL. These modules exist in all HN's, CCN's, and SN's.

NWLU-SVR is a module performing the same functions as an FCCL and may exist in an HN and a CCN. It is used to send or receive APU's from NWLU's. The macro instructions GET, PUT, OPEN, and CLOSE, used by user processes to handle networks, can be handled by this module.

NTW-MGR is a module, which is used to supervise and manage a message transmission network. It also manipurates HN's, CCN's and SN's, except for the host node in which it exists.

(4) MCAM-RAS is used to handle requests from an operator for data unit tracing, CCN/SN testing, or system program file (SPF) loading into some other node. When such a request is made, it informs the NTW-MGR that the request is being made, and then outputs the results of the operation in print form.

(5) The TOSWAY-RAS, communicating with the DLCL, collects, edits, and outputs the records of error tracing and sequence tracing by all stations, in accordance with instructions given by an operator. A hardware unit suspected of causing an error may also be tested by an operator using this program. Restoration of the physical transmission path is directed by this program. In addition, this program

supervises and manages the logical network, i.e., the data link level hardware in the PIO access network, BIO access network, and message transmission network.

Characteristics

- Since the various functions necessary for the logical network are hierarchically arranged in a layer structure, the construction of networks and the alteration of user processes in function are relatively easy.

- If part of the network fails, the influence on the whole system can be reduced to a minimum.

- User processes can perform transmissions by using identical network access macro instructions, regardless of the types of processors involved.

- User processes can readily access the resources of any node by using the same FCP.

- Any user process may determine an application protocol (APP) suitable for a system on which it is to run, by using the FCP provided by the FCCL.

- PIO devices or BIO devices may be accessed by means of the same macro instructions, without regard to whether or not they use the associated networks for the purpose. This makes high-speed processing possible.

- A processor of a different architecture can be used as an SN if a special CCS is implemented.

- Since errors occurring in the logical network are processed either by the MCAM-RAS or TOSWAY-RAS, trouble shooting can be performed immediately.

- At an HN, any system program in it, or a user process in some other HN or CCN, can be compiled or assembled. Also, down-line loading can be performed using the MCAM-RAS at an HN.

CONCLUSION

In the new ring data bus system, the centralized processing method is adopted in the PIO access network and the BIO access network to overcome deficiencies in the distributed processing method. However, distributed processing in the message transmission network has drawbacks which require further investigation. They are listed below.

- Although network component failures can be found for analysis, such failures, which may occur anywhere in the network, are difficult to find.

- The system software requires a high overhead.

- It is still necessary to combine this medium-scale network based on minicomputers with a network based on large-size computers.

Moreover, networks will become more complex and competent than at present with the future development of hardware speed and performance. The introduction of optical information transmission into actual use will accelerate this tendency. Thus, we are determined to make continuing efforts:

- To implement a diagnostic tool capable of detecting even failures of hardware unit parts.

- To replace more software-performed functions by hardware.

- To develop a network model simulation capable of analyzing the performance of the logical network.

We feel that in the near future such efforts will make it possible to develop a network which is both widespread and tightly unified.

REFERENCES

IEC SC65A/WG6 (1978) IEC Publication "PRO" 'PROWAY' Functional Requirement

ISO/TC97/SC16 (1979) Reference Model of Open Systems Interconnection (Version 4 as of June 1979)

Y. MATSUMOTO (1978) A distributed processing system and its application to industrial control. NCC, 1273-1279.

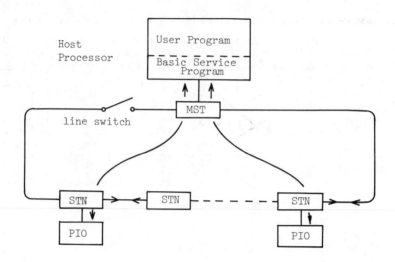

Fig. 1 Configuration of the first ring data bus system

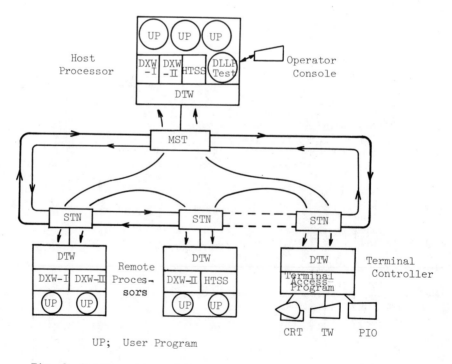

UP; User Program

Fig. 2 Configuration of the second ring data bus system

Distributed Computer System

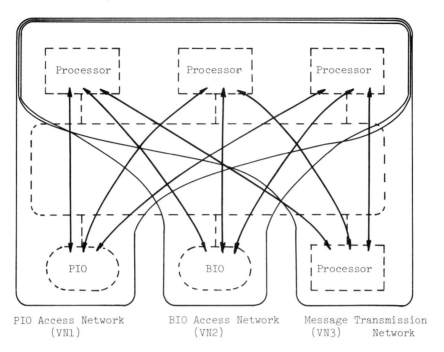

Fig. 3 Logical network

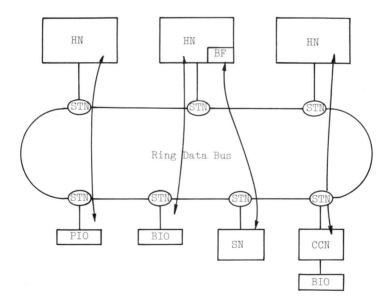

Fig. 4. Logical network nodes

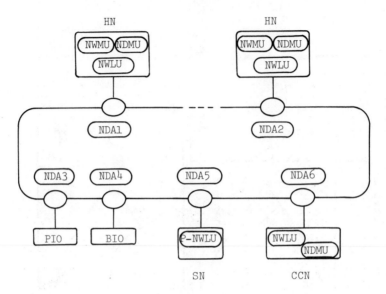

Fig. 5 Addressing system

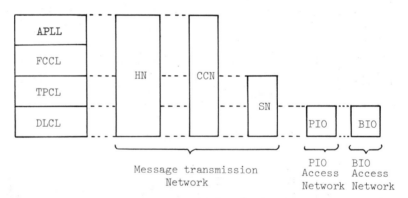

Fig. 6 Logical network structure

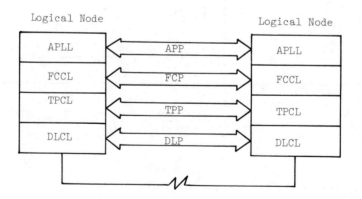

Fig. 7 Protocols

Distributed Computer System

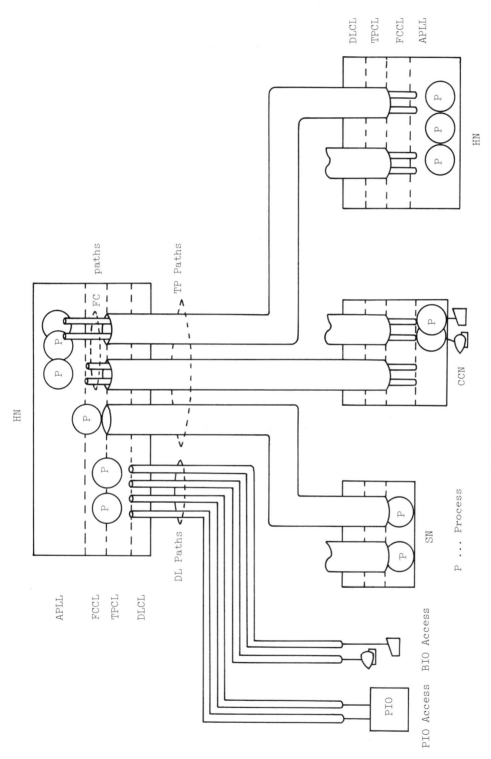

Fig. 9 Data paths

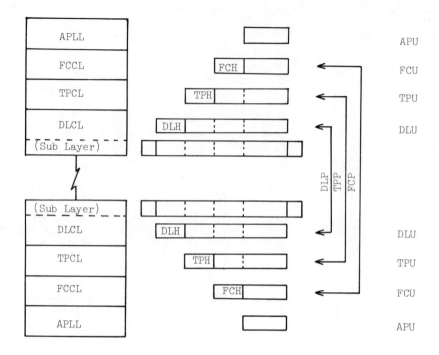

Fig. 8 Relationship between layers, protocols, and data units

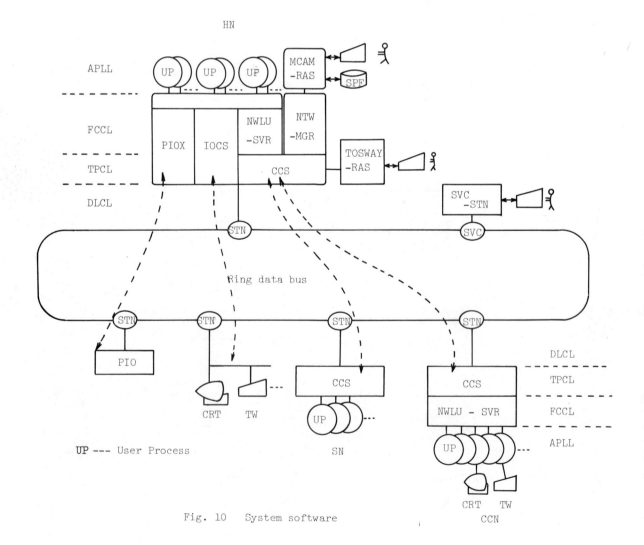

Fig. 10 System software

DISCUSSION

Willard: I would like to inquire as to your intentions in regard to downline loading from the host node to the other nodes. Do you plan to downline load tasks, or merely attributes, into the remote nodes?

Kawabata: We intend primarily to downline load user function programs; that is, application programs.

Hyde: Do you have any data on the percentage of actual traffic bandwidth which is associated with the layers of protocol? That is, if we look at the bandwidth on the line that is actually traffic, some percentage of it is protocol and some percentage of it is application data. What percentage ends up being in the protocol?

Kawabata: The line speed of the second ring data bus system is about 1.5 megabit/second, but the actual effective line speed is about 1 megabit/second. Using the new ring bus system, the line speed is about 50 percent of that. I expect that this can be speeded up, but software overhead is very, very high. In hardware products, for example, in the accelerated system, the transport control layer was developed using hardware or firmware.

Müller: You mentioned the transmission speed was 10 megabits/second. Can you give some information on the distance over which you can achieve that speed? Second, what type of cable do you use? Third, what type of coupling do you use at a substation? And, fourth, did you consider fiber optics?

Kawabata: We have executed a system using optical fiberglass already. As to the distance, it is about 100 kilometers total, but for two stations it is about 3 to 5 kilometers. The bus system uses coaxial cable, but the first and second ring data bus systems use twisted pair. Since I am a software engineer, I do not know what type of coupling is used at the receiver stations.

Gellie: Is this a true loop system?

Gellie: Is this a true loop system? That is, does all traffic flow around the loop in the same direction?

Kawabata: Yes, it normally flows in the same direction. However, if a station fails, the traffic flows bidirectionally, using the master ring data bus and the standby ring data bus.

Sandmayr: You divided the logical network into three virtual networks. Can you explain some reasons why you chose these partitions and did not, for example, consider business input-output as a message?

Kawabata: The reason is performance. In the PIO, BIO access networks, we can access in about 1-2 milliseconds.

Wilhelm: With reference to Figures #7 and #8, if two application functions reside in the same processor (computer), is it possible to bypass the lower protocol levels, going directly between APLL's or between FCCL's, without involving TPCL and DLCL?

Kawabata: We provide special macro instructions for communicating in the same processors.

TRANSACTION PROCESSING IN DISTRIBUTED CONTROL SYSTEMS

R. G. Wilhelm, Jr.

AccuRay Corporation, Columbus, Ohio, U.S.A.

Abstract. While the flexibility of a process control system may be enhanced by modular design, the effectiveness of the system often depends upon coordination of the modules' actions. The conflict between designing clearly bounded modules and providing for coupling between them gives rise to important design tradeoffs in two major areas: the appropriate size and scope of modules for the application, and the amount of knowledge any one module must have about any other module for communication and coordination. This paper discusses a general approach to coordination and communication between modules with specific examples concerning how interdependent control functions may activate and deactivate each other. In addition, the paper describes methods for on-line reconfiguration of the relationships between modules.

Keywords. Computer control; computer programming; distributed control; distributed intelligence; hierarchical systems.

INTRODUCTION

Much attention has been directed recently toward the design of data buses and communication protocols for distributed computer control systems. While these are essential to the operation of such systems, they have a tendency to divert our attention from the objective: process control. Further, there is a temptation to allow the hardware structure and communication techniques dictate the structure of the control software - a classic violation of the venerable dictum that form should follow function. From the control engineer's viewpoint, the hardware configuration of a distributed computer control system might ideally form an image of the control functional block diagram. There are many other practical forces which influence the hardware configuration, however, including location of sensors and actuators, location and form of operator interface, limitations of processor capacity, availability of special purpose processors to speed certain tasks, bus limitations, reliability, and many others. In order to optimize the flexibility and effectiveness of the control system, the control software should make maximum use of the hardware without having its own structure dictated by the same extraneous forces.

The primary factor influencing the structure of the control software should be the nature of the control problem itself. Other important shaping forces, however, include ease of implementation, understandability, ease of modification, and reusability for similar applications. These factors all point toward a modular design approach to control software. The challenges in achieving modularity arise from the need to make successful tradeoffs on pairs of apparently conflicting design objectives. The most important tradeoffs fall into two categories:

1. Design of the modules themselves. The software modules should be specifically useful for solving the control tasks for the given application without being so specialized as to prohibit multiple usage.

2. Modules' awareness of each other. Although modules must communicate with each other in order to form an effective system, each module should be unaware of the internal structure of other modules to the maximum extent possible.

This paper is concerned with methods for effectively making the tradeoffs in category 2.

If communication is to occur between modular functions which do not have intimate knowledge of each other's internal structure, then they must agree upon the language to be used for their communication. It would be too presumptuous to attempt the definition of a single language to serve all applications. Indeed, it is the author's conviction that the language used for communication between high level control functions should be specifically designed for the application. It is not too much to hope, however, that a general framework for communications can be developed. The following sections introduce such a framework, discuss its attributes, and give examples of a particular language for auto/manual logic.

CONTROL COMMUNICATIONS FRAMEWORK

Character of Control Communications

Consider the general problem of communication between two control elements. The communications required are likely to fall into one of three categories:

1. Petition - One control element requests another element to take some action.

2. Inquire - One element requests some data from another element.

3. Inform - One element transmits unsolicited data to another element. (By unsolicited we mean only that the information is not the response to a specific inquiry).

By choosing the names "petition", "inquire", and "inform" we are purposely avoiding the more specific "request task", "fetch" and "store" which have special meanings in computer applications and which may be but single examples of a petition, inquire, or inform, respectively.

To categorize the content of the communications requires a detailed knowledge of the application. In fact, it is not necessary to know the content in order to define communication techniques. It is much more useful to consider the relationship between a control element and the information required by it. In particular, we note that while some information is requested or transmitted by an element in the process of performing its function, other information actually causes certain tasks of the element to execute. This duality of the causal role between tasks and information is a fundamental quality dictating an appropriate communication sequence.

Subscription and Notification Mechanisms

In a distributed system if one frequently executing element requires several pieces of information from other elements, it would be grossly inefficient to inquire for each piece of data at the point in the program where it is used. Conversely, if some information is used only rarely by a foreign element, it is inefficient to routinely transmit it when it is not required. A useful technique for reducing unnecessary communication is to allow elements to enter and cancel "subscriptions" with other elements for certain information.

Subscriptions for data on a periodic basis have obvious application in real-time control. Depending upon the sophistication of the system the subscriber might be able to specify the frequency at which he wishes to receive the data. The responsibility then lies with the source element to keep an up-to-date subscription list and transmit the data to the subscribers at the appropriate time. The ability to enter and cancel subscriptions in real time permits the most efficient use of communication lines, and is especially useful if the control objectives change during normal operation.

Another valuable use for subscriptions, however, is to request a receipt of notification of the occurrence of a random event. For example, if a higher level control element requires the services of a DDC loop, the higher level element should receive notification if the DDC loop turns off for any reason. The notification mechanism saves the continual transmission of information which changes very infrequently. Furthermore, by cancelling the subscription, the subscriber can avoid the requirement for complex logic to determine whether the notification is of any interest at present.

Structure of the Communication Facilities

We now undertake to define the structural aspects of a communication system suited to high level control functions. For simplicity let us temporarily ignore the distributed nature of the system and concentrate on the modularity of the control functions. Perhaps the simplest useful view of communication between elements is given in Fig. 1. We may term an individual act of communication a "transaction". The transaction must be initiated by one element (which we call the "subject" element) transmitting a petition, inquiry, or information to an "object" element. There may or may not be a reply expected from the object, depending upon the particular transaction. This view implies not only that the two elements share a common language for communication, but that each has enough self-contained intelligence to interpret and respond to the information received.

We now wish to refine our view of a transaction. The first stage of refinement is to recognize that while the object element must have the intelligence to respond to communications from other elements, this intelligence may be separated from any independent algorithms required for the control function. The point is illustrated in Fig. 2. For example, suppose that the object element, B, is a DDC loop whose algorithm executes periodically, say once per second, to compute actuator adjustments. The communication processors will, on the other hand, only execute when called upon, and may perform such functions as turning the DDC loop on (including the attendant initialization procedures) or updating the setpoint, subject to limit checks. Note, however, that communications to other functions from function B could logically emanate from either function B's algorithm or its communication processors.

Further refinement of the communication structure follows from recognition that a transaction consists of two distinct phases. The handling of the actual communication, involving translation of messages into appropriate protocols, searching directories for addresses, and other overhead functions are particular to the type of transaction - not to the elements involved. The response to the given transaction, on the other hand, may be highly individualized for a particular element. For example, if the subject element petitions another element to turn on, the steps taken to do so may be unique to a particular object element. It is wise to provide system utilities to handle the functions which are common to all users of a transaction.

Figure 3 illustrates a structure which supports all of the facilities we have described. The Transaction Processor is a system utility which is simply a collection of subroutines which can be called by any element. The subject element calls the routine for the particular transaction it desires, passing its own identity and that of the object element. The Transaction Processor issues one or more commands to the object, whose Command Processors may be unique to that element. The responses are analyzed by the Transaction Processor which evaluates them and returns a reply to the subject. Some of these steps or components may be eliminated for simpler transactions, such as when no reply is required.

It is also convenient to allow the Transaction Processors direct access to an Element Info Table in each element's data base, containing certain information in a standardized format. This information may include pointers to the various Command Processors, subscription lists, etc. Some transactions may be defined just to enter and cancel subscriptions. Other transactions may send notices or data to the subscribers, instead of being directed to a single object element.

Communication Among Distributed Control Functions

The structure developed in the previous section is well suited for extension to distributed control, since the Transaction Processor can decouple the control elements themselves from the communication protocols. The Transaction Processor can assume the task of encoding and decoding messages and can interface with the systems' communication utilities.

Figure 4 illustrates the arrangement for communicating between control functions located in different computers. Transaction Processor 1 must recognize that the object element resides in another computer. This fact may be encoded in the ID of the object element or may become apparent only after referring to directories. TP1 then simply relays the transaction to TP2. TP2 does all of the work, interfacing with object element just as though the transaction had been initiated within Computer 2. When finished, it returns the reply, which is simply relayed through TP1 to the subject element.

EXAMPLE APPLICATION: AUTO/MANUAL LOGIC

The power of the communication structure and techniques discussed thus far can be illustrated by applying them to the coordination of the status of individual but interdependent control functions in a complex modular system. In many applications, an elements status may be adequately defined (at least insofar as other elements are concerned) by three binary state variables: ON/OFF, ACTIVE/INACTIVE, AVAILABLE/UNAVAILABLE. ON/OFF is rather self explanatory. When an element is ON we may wish to modify its behavior by making it INACTIVE at certain times. When an element is OFF, it may or may not be AVAILABLE to be turned on depending upon external factors such as local/remote safety switches, interlocks with other elements or simply the AVAILABILITY of other elements upon which it depends.

We can make full effective use of the transaction processing/command processing approach defined previously be employing the following strategy:

1. The auto/manual logic is separated from the control algorithm for each element and is only executed on demand.

2. The auto/manual logic for each element determines the status of that element, which governs the behavior of the algorithm.

3. The auto/manual logic for each element is responsible for responding to a standard set of commands for state transitions. The way in which an element responds to the commands may be unique.

4. Standard transactions are implemented as system utilities to facilitate auto/manual interactions among the elements in a cooperative manner.

5. The subscription/notification technique is used to inform elements of status changes in other elements.

The particular realization described in the following subsections has been in use for some time by the author and his colleagues.

Auto/Manual Command Processors

Each element must respond, possibly in a unique way, to commands to change state. Given the definitions of the three binary state variables ON/OFF, ACTIVE/INACTIVE, AVAILABLE/UNAVAILABLE, logical choices for the names of appropriate pairs of commands would be: START (turn ON), STOP (turn OFF);

ACTIVATE(become ACTIVE), DEACTIVATE (become INACTIVE); LOCK (become UNAVAILABLE), and UNLOCK (become AVAILABLE). In responding to these commands in its own unique way, an element might need to influence the states of other elements, perform some initialization functions, and so on. Since the command processors execute only upon demand we must also respond to NOTICE commands from other elements to which we have subscribed. The command processor must interpret the notices which come as a consequence of status changes and determine what this element must do as a result.

Auto/Manual Transaction Processors

The choice of transactions available to the elements determines the ability to coordinate the elements in a highly interactive system.

Transactions for subscription and notification. First we provide two transactions to support the subscription/notification mechanism. SUBSCRIBE and CANCEL SUBSCRIPTION. These allow each element to request notification of status changes from any other element or to terminate that arrangement. A NOTIFY transaction may be defined, but rather than require the element's algorithm or command processors to invoke a NOTIFY, we may simply have the other transaction processors perform that function whenever a status change is caused by them.

Transactions for transitions between ON and OFF. When one element wishes another to be on it may issue a REQUEST. If the object element is already on, no commands are issued; if not, a START command is issued. The REQUEST transaction processor automatically enters a subscription for the subject element as well. If the subject element no longer requires the object to be ON, he may WITHDRAW his request (another transaction). The WITHDRAW transaction processor will not issue a STOP command unless all subscribers have withdrawn their requests. If the subject element should CANCEL his subscription that transaction processor automatically withdraws his reqeust. If the subject insists that the object should turn off he may issue an ABORT transaction. Of course, all subscribers will be notified if a status change results.

Transactions for transitions between ACTIVE and INACTIVE. If an element wishes another to become INACTIVE, he may issue a SUSPEND transaction. This act automatically provides a subscription also. If the subject no longer requires the object to be INACTIVE he may use the ALLOW transaction. The transaction processor will not issue an ACTIVATE command as long as any subscriber has suspended and not subsequently allowed the object. Cancellation of a subscription automatically generates an ALLOW transaction for the subject element.

Transactions for transitions between AVAILABLE and UNAVAILABLE. If an element wishes to LOCK another element he may issue a DISABLE transaction. When he no longer requires the object to be unavailable he may issue an ENABLE, but the transaction processor will not issue an UNLOCK command until all subscribers who have disabled the object subsequently enable it. Cancellation of a subscription automatically enables the object for the subject element.

Comments on the auto/manual logic. Table 1 summarizes the relationships among the transactions, commands, and forms of notification. In addition, there is a transaction called CHECK which simply inquires as to the current status of the object element. It should be apparent that the methods described provide a very cooperative environment so that the actions of one element impose minimal adversity upon other elements. The methodology and rationale are discussed in more detail by Wilhelm (1979a).

STATE	COMMANDS	NOTIFICATION	TRANSACTIONS
ON, OFF	START STOP	STARTED STOPPED	REQUEST WITHDRAW REQUEST ABORT
ACTIVE, INACTIVE	ACTIVATE DEACTIVATE	ACTIVATED DEACTIVATED	ALLOW SUSPEND
AVAILABLE, UNAVAIL.	LOCK UNLOCK	LOCKED UNLOCKED	DISABLE ENABLE
———	———	———	CHECK
———	NOTICE	———	SUBSCRIBE CANCEL SUBSCRIPTION

TABLE 1 - Summary of Transactions, Commands, and Forms of Notification for Auto/Manual Logic

ON-LINE RECONFIGURATION

Thus far in our discussion we have dealt with control systems of a preprogrammed structure. Although we have provided mechanisms (e.g., subscriptions) for elements to establish and sever communication connections with each other, the logic dictating these changes was assumed to be preprogrammed intelligence in the elements. There are some applications, however, wherein the relationships between elements must be reconfigured on-line. Although experimental or development systems fall into this category, we are more concerned here with a system applied to a process which itself is reconfigured during normal operation. Consider, for example a batch or continuous blending process consisting of several chemical storage tanks whose outlet streams can be directed into any one of several mixing vessels. The blending control system will need to associate flow controllers with mixing tank level controls and product blending recipes in accordance with the products to be made at a given time. Such a system may receive its configuration directives from operator-entered data, from contact inputs from the process itself, or from other sources.

Defining the configuration

Generally the structure of the control system is known, in the sense that the designer knows that certain types of control elements will influence the behavior of other types of elements in a fixed way, but the particular relationships are not established. Such a system requires utilities for establishing and modifying these relationships on-line, and a record-keeping system. Let us consider what information is required to define a relationship. First, the existence of a linkage between two elements must be known. Second, the nature of the relationship must be established. For example, in a hierarchical control system we have "leaders" and "followers". Note, however, that the relationship is relative, since a given element may be a follower in relation to one element but a leader of several others. Finally when several elements bear the same relationship to another one, they may still perform different "roles", in much the same way as individuals reporting to the same supervisor may be assigned distinctly different tasks to perform. Figure 5 illustrates a hierarchical arrangement of elements which may be divided into groups, where each group contains a leader and one or more followers, and each follower may be assigned a role.

Configuration records. A convenient and modular means of recording the configuration is to provide each element with an "association list" as a part of the Element Info Tables mentioned earlier. In this list is recorded the identity of each element which has been linked to this one for some control function, the relationship of the other element (leader, follower, etc.) and its role. In hierarchical systems it is convenient to divide the association list into two parts: a leadership list and a followership list.

Configuration changes. The utilities for establishing and changing the configuration records may be implemented as transaction processors which may either be granted direct access to the association lists or interface with command processors in each element which make the table entries. The required transactions are to create a relationship, destroy a relationship, or modify the role of an element in an existing relationship. Note that the creation of any relationship involves complementary entries in the association lists of at least two elements.

Program References to Unknown Elements

In the earlier discussion of communication transactions where the system configuration was fixed, an element's programs could contain references to specific elements. In configurable systems these references must be made indirectly via the association lists. Furthermore, unless specific places in the association lists are reserved for elements with particular relationships and roles (not a completely flexible solution), it will be necessary to search the lists for the appropriate elements. An excellent method for handling these indirect references is a language statement similar to the DO statement in FORTRAN of other languages. The "selective DO" must be capable of selecting a subset of the elements in the association list. Thus "do loops" may be established to execute any arbitrary set of instructions, including communication transactions, once for each element bearing a particular relationship and role to the element being executed. The syntax of such a statement would be:

DO LABEL RELATION_SPEC ROLE_SPEC ID RELATION ROLE

followed by any arbitrary program statements to be executed in the do-loop, the last of which has the specified LABEL. The RELATION_SPEC and ROLE_SPEC specify what subset of elements is to be considered in this do-loop. The DO statement returns the actual ID of each element, one for each execution of the loop, and the actual RELATION and ROLE of that element.

More discussion on this technique is found in Wilhelm (1979b), with reference to a process control example problem.

CONCLUDING REMARKS

The techniques discussed in this paper have been implemented by the author and his colleagues and are routinely used in control systems for the paper and other sheet processing industries. The auto/manual logic package, called Transactional Auto Logic, is used in all AccuRay 1180 Micro® systems, while the on-line reconfiguration tools (designated MAP for Multiloop Application Package) are used for applications requiring that feature, such as blending controls and refiner controls, where the customer's process is highly configurable.

These techniques encourage modular design from the control function viewpoint and aid in overlaying the control structure upon the architecture of a distributed system. Of course, they do not alleviate the necessity to consider the affects of communication delays upon the control system, but rather provide effective ways of living with that inconvenience via the subscription and notification mechanisms.

REFERENCES

Färber, G. (1978). Principles and applications of decentralized process control computer systems. Proc. IFAC Seventh Triennial World Congress, 1, 385-392

Gaspart, P., and P. Vandenbussche (1978). An ergonomic and reliable system for distributed process control. Proc. IFAC Seventh Triennial World Congress, 1, 715-722.

Schneider, G.M. (1979). Computer network protocols: a hierarchical viewpoint. Computer, 12, Sept., 8-10.

Wilhelm, R.G. (1979). A methodology for auto/manual logic in a computer controlled process. IEEE Trans. Autom. Control, 24, Feb., 27-35.

Wilhelm, R.G. (1979). On-line coordination and configuration in modular process control systems. Proc. 1979 JACC, Denver, AIChE, pp. 78-83.

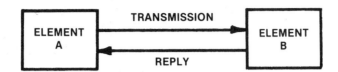

Fig. 1. Simplest view of communication between elements.

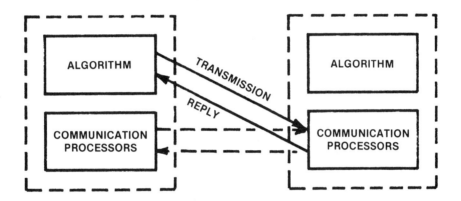

Fig. 2. Separation of communication functions within each control element.

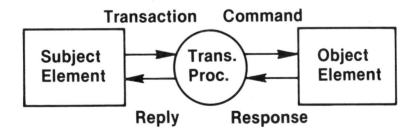

Fig. 3. Using Transaction Processors to perform communication functions common to all elements.

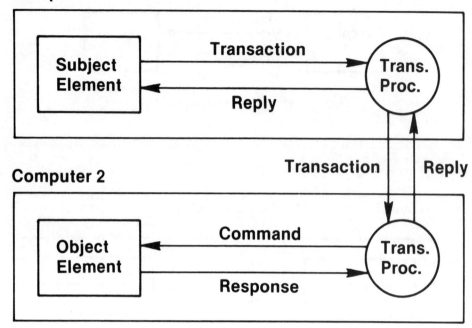

Fig. 4. Using Transaction Processors in a distributed system.

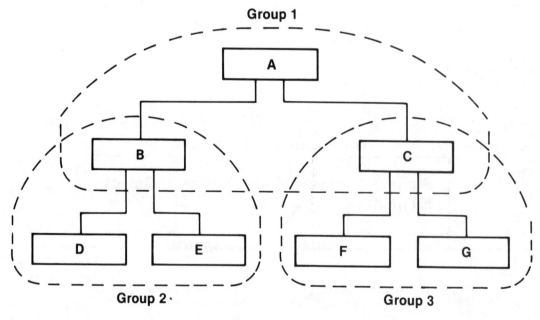

Fig. 5. A hierarchical structure of elements subdivided into groups.

DISCUSSION

Miller: Do I understand that all elements, as you define them, communicate only by messages and don't require concepts such as shared memory?

Wilhelm: That's correct, but bear in mind that I am talking about specific applications. For example, I treat automanual logic as a specific application, and in our systems we have implemented the automanual logic in the sense that I have described it. If any element wishes to interface with another element for automanual logic, it must do it through this mechanism. There may be other mechanisms, but I think it is highly undesirable to use shared memory type approaches if you really want to maintain modularity.

Miller: What is your definition of an element and how is it addressed? Is it always referred to symbolically at the highest level of addressing?

Wilhelm: I've defined an element simply to be a control function which we wish to treat as a separate module. For example, each DDC loop in the system is an element and each higher level control loop is an element. Another element in the system, for example, is a set of logic which looks at what the operator has done at the operator station, such as pressing push buttons and so forth, interprets that information, and then issues the necessary transactions to turn on or turn off control functions in the system. Each element in the system in our implementation is given an identification number, which is usually broken down into two parts: one is a type number (for example, all DDC loops which share the same algorithm subroutines have a certain type number that they share) and the other part is the unique identification number of that element. It's a numerical code; however, it could as well be a symbolic code.

Miller: What is your definition of a transaction? How does it relate to the data base and how is it different from a message? Is it composed of messages?

Wilhelm: In the sense that I am using it, one could think of a transaction as a message itself; however, since I am trying to define it very carefully at a higher level, at the control level rather than at the communication network level, I prefer to think of it as something a little different from a message. A transaction is defined to perform certain functions in the control system. I have defined a set of transactions which are useful for automanual logic. Those transactions are indeed messages which are sent to the transaction processor, which then acts upon them to influence the behavior of another element in the system. However, they are defined for this specific application, and I have a different set of transactions which I use in the multiloop reconfigurable application. In that case, the transactions allow you to do things like link and unlink functions, create and destroy these linkages, or change and modify the roles of particular elements. Specifically, there is a transaction called "modify role' which might tell the element that it is no longer the first priority load in this system, but that it is now the second priority load.

Willard: I'd like to explore your concept of subscriptions since it sounds like a very useful mechanism for reducing the amount of interprocessor communications. By analogy, if you subscribe to a newspaper, either the sink or the source has the right to cancel subscription. I'm wondering if you have a mechanism by which the source of data in a subscription can cancel it and how does the sink get notified?

Wilhelm: We have not provided that mechanism in our system in any of the applications of subscriptions so far. I can visualize a few cases where that might be useful. I think that the final act, if you are going to cancel a subscription, would have to be to send one last notice to the subscriber, saying, "You are no longer a subscriber, so don't expect any more messages from me." I think that would be fairly simple to implement. We have exclusively used the subscription mechanism for notification. We have not, thus far, used it in a system to send periodic data. However, in the paper I described that and I do think it's a useful mechanism.

Willard: If a sink requests that to be added to a subscriber's list and the subscriber's list, which is then the source, is already full, can you cope with that situation?

Wilhelm: Yes, there are a number of conditions in the system which generate errors. For example, it may be that you wish to subscribe to an element that doesn't exist in the system, so we've provided some failure modes which are failsafe. Typically, those error conditions only occur during an initial configuration and check out of the system and you don't find any of those errors occurring in real time. However, we've tried to design the system so that the failure is soft and if you receive a notice that says that an element doesn't exist, the rest of the software doesn't depend on sending setpoints to that element or receiving data from it; that is, you respond to the fact that you have a configuration error. For example, when you say "DO for all of my leaders," if the DO function comes back and says I can't even find a list for you, you don't seem to be a viable entity in the system, or I can't find any leaders for you, that represents an error and the system then responds to that.

Müller: You have given an excellent example for three state variables for your application. For your practical implementation, could you give some figures about the time it takes, in addition to the communication that takes place, for the transaction processor to do the job? Can you also give some ideas about limitations of the approach for more complex things, where you need many more than three state variables?

Wilhelm: Dealing with the second question first, it's awfully hard to give any kind of quantitative answers. Certainly, more variables will complicate the definitions of transactions and commands, and the relationships between them. Basically, the complications arise when one is defining the degree of cooperation. As you've seen, I have more transactions defined than commands, in order to provide the necessary cooperation for auto-manual logic. For example, I have only two commands to turn a function on or off, but I have at least three transactions: request, withdraw request, and abort. In addition, whenever one does something like enter a request, he automatically gets a subscription entered in our implementation. Thus, he will receive notices if that object element aborts. All of those things complicate the issue and to say how much is a little difficult. However, I think you have those complications to deal with if you are going to provide that kind of cooperative arrangement in any system. Therefore, I don't think the particular mechanism that I have described makes it any harder.

In terms of timing and overhead, I can't give you any figures on timing, although I can give you some figures on memory. The typical command processors for the auto-manual logic only take a few hundred words each in a 16-bit minicomputer. Again, however, many of the elements share the same command processors. For example, all DDC loops respond the same way to commands, so they all share the same command processors for start, stop, activate, and deactivate. The central transaction processors for handling all of the auto-manual transactions take about 2K of memory, so they are also relatively simple to implement.

Humphrey: I was very interested in the discussion you had about roles, leaders and followers. Have you addressed the problem of priorities in such a structure?

Wilhelm: Only in the sense of control priorities. In the example that I gave of power demand control, one has a series of electrical loads in the process plant which may be shed in order to avoid exceeding power demand limits and getting excessive energy charges. The priorities of those elements, in the control sense (which ones can you turn off first, which ones can you turn off second, etc.), are handled by the internal software in a pre-programmed manner. All of these routines must, of course, be reentrant and recursive, because any one transaction may generate the same transaction directed toward another element.

Humphrey: I was specifically thinking about a situation where I have two or three different functions that want to establish a role with a certain element. Have you set up something to handle those kinds of situations?

Wilhelm: Not at this point of time. Typically, the roles are established by the operator entry in the applications we have dealt with so far, rather than being automatically generated within the system. An operator, for example, establishes whether certain flows are to be held constant, or are to be ratioed to the total, or what their role is to be in a blending system. Similarly, the operator establishes the priority in which loads are to be shed in the power demand system. We've applied this to about three applications so far. These are a blending control system, refining control in a paper mill where refiners can actually be piped from one place to another, and the control of multiple gas dryers on the paper machine where you wish to regulate different dryers at different points in time, or to ratio certain dryers to a certain output.

PARALLEL PROCESSING CONCEPTS FOR DISTRIBUTED COMPUTER CONTROL SYSTEMS

D. AlDabass

Control Systems Centre, University of Manchester Institute of Science and Technology, Manchester, U.K.

Abstract   It is proposed to utilize the spare processing power embedded in a real-time computer control network for the computation of a common distributed task in a co-ordinated manner. The algorithm of Hierarchical System Theory is used as a medium to explore the principles involved in time-sharing this distributed background job with the computation of the real-time control algorithm. An interconnection structure is put forward to satisfy the resulting data-processing requirements.

Keywords.   Real-time systems; distributed computer control; time-sharing; networks; multiprocessor interconnection schemes; hierarchical system theory; optimization.

1. INTRODUCTION

The motivations for Distributed Computer Control Systems have included combinations of Technological, Functional and Economic reasons. Economic reasons stemmed from the communication cost of sensing and actuating relatively remote parts of the plant; functional reasons were motivated by the relative independence of subsystems; and finally, technology provided an attractive solution through the use of inexpensive local processing units (Down and co-workers, 1976; Gentler, 1975; Blanc and others, 1976; Jones, 1977; Schwartz, 1978).

In an attempt to find methodical procedures for structuring and studying these systems, decomposition techniques and hierarchical optimization algorithms have been used extensively (Bauman, 1968; Macko, 1966; Mesarovic, 1970; Nicholson, 1978; Sage, 1968; Singh,1977; Singh and others, 1978; Takahara and others, 1969; Wismer, 1971). These methods have the advantage of satisfying two major objectives in the overall problem. The first is that the plant, and its model, usually exhibit distributed structures amenable to decomposition; and secondly, the growth in the theory of hierarchical optimization provided sound analytical foundations for their description and control. In this respect HST (Hierarchical Systems Theory) can be employed in two ways. First, it can be used as an analytical tool during the design stage to compute the trajectory of each subsystem to ensure the optimum performance of the total plant. This computation can be typically carried out on a single processor, even though the analysis implies a decomposition on a multiprocessor system. The advantage being that it is easier (for a single processor) to optimise, say, 10 systems of 5th order, rather than one system of 50th order. The resulting trajectories and/or their corresponding control sequences, are then loaded into the individual controllers for each sub-system of the plant. In this case the HST algorithm is computed sequentially, and then distributed among the plant controllers as determined in the analysis.

In the second method, the computation of the HST algorithm itself can be distributed among a number of processors. Three methods of distribution can be considered here depending on the relative subsystem order allocated to the infimals. The simplest method gives a direct correspondence between the infimal and its processor, i.e. the whole of subsystem i is optimized by processor i. The second method leans more to the sequential case and allocates more than one subsystem to a given processor. The practical significance of this method is that some subsystems are by their physical nature small order, and do not need a whole processor to compute their trajectories. The third method complements this argument by dividing the arithmetic workload of a relatively large subsystem among a number of processors to even out the computations on all processors.

In the general case a distinction must be drawn between the processors that compute the optimal trajectories, and those that carry out the control. In off-line applications, the first group are usually housed within a large computing system available to the design team as a computing aid; while the second group form a collection of microprocessors distributed throughout the actual plant to be controlled. In Real-Time applications, these two groups can be mapped onto the same physical collection of processors, but multiplexed in time. In this mode, the processors alternate between the two distinct roles of operation; computing the trajectories first, during which time they cooperate according to one of the three methods described above, and then using these trajectories perform the control of the plant subsystems.

This paper aims to explore these concepts in more detail. Section two is a brief and introductory review to hierarchical systems theory and geared towards the explanation of basic concepts rather than mathematical rigor. Section three continues this theme and examines the computational aspects of the algorithm as viewed from the multiprocessor stand. Section four deals with the distributed computer control problem. The general design features involved in these systems are explored, and an interconnection scheme is put forward to meet the data processing requirements of the proposed method of operation.

2. HIERARCHICAL SYSTEMS THEORY

Consider the general nonlinear and/or time-varying system

$$\dot{x} = f[x(t), u(t), t]$$
$$x(t_o) = x_o; \quad x \in R^n; \quad u \in R^m \quad (1)$$

where n is large enough to cause computational difficulties. The problem is to generate the state (and often the control) trajectories for the time interval $t_o < t < t_f$ which minimise the performance measure

$$J = \Theta[x(t_f), t_f] + \int_{t_o}^{t_f} \phi[x(t), u(t), t] dt \quad (2)$$

2.1 Solution Without Decomposition

The conventional method of solution is by forming an adjoint system given by the equation

$$\dot{\lambda} = -(\frac{\partial f}{\partial x})^T \lambda - (\frac{\partial \phi}{\partial x})^T \quad (3)$$

whose solution for $t_o < t < t_f$ must satisfy the end condition

$$\lambda(t_f) = (\frac{\partial \Theta}{\partial x}(t_f))^T \quad (4)$$

This gives rise to the well known two-point boundary value problem (TPBVP) of optimal control. The control trajectory associated with the optimal solution is given as a side product by the algebraic equation

$$(\frac{\partial f}{\partial u})^T \lambda + (\frac{\partial \phi}{\partial u})^T = 0 \quad (5)$$

(i) <u>Analytical solution</u>: This is possible only when the following conditions are met: (a) f and ϕ are both differentiable wrt both x and u to give an analytical expression for both λ and u; (b) the form of these expressions, and that of f after substituting u, should be such that an analytical solution for both λ and x (as functions of each other and t) can be found; (c) Θ should be differentiable wrt x to give an expression for $\lambda(t_f)$; and (d) the resulting set of algebraic equations for $\lambda(t_o)$ and $x(t_f)$ are solvable analytically.

(ii) <u>Numerical solution</u>. In general, concitions (a) and (c) can be much more easily met than (b) and (d). A popular method of solution is to integrate x and λ by a suitable simulation technique using a given value for $\lambda(t_o)$. This reduces the solution to a static iterative, optimization process to find the value of $\lambda(t_o)$ which satisfies the end condition of equation (4). In effect a search in the n-dimensional $\lambda(t_o)$ space is conducted, which usually gives a computing time proportional to n raised to the power of a,

$$T = \tau \cdot n^a$$

where τ = time to generate one complete solution for a single (vector) value of $\lambda(t_o)$; and a = constant related to the specific optimization technique, of the order of 2; n^a is the total number of iterations required for the search.

2.2 Hierarchical optimization.
n^a can be very large even for systems of relatively small order. Consider decomposing the system into N smaller systems each of order n_i and having m_i control variables,

$$\sum_{i=1}^{N} n_i = n; \quad \sum_{i=1}^{N} m_i = m \quad (6)$$

and

$$\dot{x}_i = f_i[x_i, \pi_i, u_i, t] \quad (7)$$
$$x_i(t_o) = x_{io}; \quad x_i \in R^{n_i}; \quad u_i \in R^{m_i}$$

The elements of π_i are the state and control variables which appear in f_i apart from x_i and u_i. They are the interconnection constraints and formally expressed as

$$\pi_i = g_i(x_j, u_j), \quad j \neq i$$

where g_i is a 'selection' function and reflects the appearance of the other subsystem's state and control variables in f_i. This function can be more accurately viewed as a collection of functions, each one representing the appearance of the variables of one other subsystem j in the ith unit,

$$\pi_i = \sum_{j \neq i}^{N} g_{ij}(x_j, u_j) \quad (8)$$

where g_{ij} simply determines which elements of the (x_j, u_j) vector variables appear in the ith unit, as dictated by the exact form of f_i.

(i) _Infimal Performance Measure._ To ensure that these interconnection constraints are satisfied during the decomposed optimization process, the performance measure of each subsystem is modified to include a new term. This in effect penalizes errors between the actual values of x_j, u_j appearing in f_i, and their predicted trajectories supplied by the supremal. However, as x_j, u_j are not available to the ith unit during its optimization process, an estimate of these variables is used. This estimate is based on the current state and control variables of the ith unit mapped onto the jth variables through the interconnection functions between the ith and jth subsystems[1]. The performance measure for the ith unit then becomes

$$J_i = \Theta_i + \int_{t_o}^{t_f} [\phi_i + \beta_i' \pi_i - \sum_{j \neq i}^{N} \beta_j' g_{ji}(x_i, u_i)] dt \quad (9)$$

where the elements of β_i relate the significance of the errors in each element of π_i to the total performance measure; e.g. errors from strongly coupled subsystems will naturally require large βs to expose small errors; while those from loosely coupled subsystems need only small βs to keep the significance of large errors down to a level compatible with the terms from other subsystems. Note that there is one element of π (scalar trajectory) associated with each state and control element in the whole system. These supplied predictions are updated by the supremal after each complete iteration by the infimals, and ultimately converge to the optimal state and control trajectories. There is also one β variable for each x and u element. The number of β trajectories may at first seem to be not only a function of the total state and control dimensions, but also the frequency of their appearance throughout the RHS; e.g. if x_3 appears in subsystems 6 and 8 then two distinct βs must be allocated to it as it cannot be assumed that the degree of coupling (and its relative effect on the performance measure) is identical for both systems. However, it turns out (see Macko, 1966) that for the decomposed problem to give the same solution as the original problem, the βs will have to play the additional role of weighting the portions of the performance measure generated by each infimal. This neatly reduces the total number of β elements to the number of π elements (n+m), i.e. for every element of x and u there is a Model and Goal coordination variables π and β respectively.

(ii) _Supremal Workload._ The problem can only be considered solved when two conditions are satisfied: (a) the computed state and control trajectories turn out to be the same as their predicted functions π, for a given (vector) function β; and (b) a value for β (vector function) can be found to optimize the sum of the infimals performance measures,

$$J(\beta) = \sum_{i=1}^{N} J_i(u_i, \pi_i, \beta) \quad (10)$$

This optimization process is shown by Pearson, (see Wismer, 1971), to be a _maximization_ of $J(\beta)$ wrt β, in contrast to _minimising_ J_i wrt u_i, see footnote[2].

The supremal therefore carries out two nested optimization processes: the outer one searches in the β function space to maximize equation (10), and the inner one searches the state and control trajectory space to satisfy the interconnection constraints. A variety of techniques are available for updating π and β between iterations, (see Singh, 1977, 1978, and Smith and Sage, 1973). One of the simplest and perhaps most effective for updating π is the 'equality updating' method, where the latest functions for x and u are used in π for the next iteration.

$$\pi_i^{k+1}(t) = g[x_j^k(t), u_j^k(t)] \quad j \neq i \quad (11)$$

The expression for β given by this method is rather more complicated, but often reduces to simple functions of the adjoint variables λ,

$$\sum_{j \neq i}^{N} \frac{\partial}{\partial x_i} [f_j'(x_j, u_j, g_j(x_i, u_i)] \lambda_j$$
$$= -\sum_{j \neq i}^{N} \frac{\partial}{\partial x_i} [g_{ji}'(x_i, u_i)] \beta_j \quad (12)$$

A gradient technique has also been used for β (see Bauman, 1968), but has been known to slow convergence,

$$\beta_j^{k+1}(t) = \beta_j^k(t) + K_{\beta_j}^k \frac{\partial H}{\partial \beta_j}, \quad K_{\beta_j}^k > 0 \quad (13)$$

3. COMPUTATIONAL ASPECTS

The decomposition procedure associated with hierarchical system theory produces algorithms which are amenable to computation both on a single processor system and a multiple processor system. The time gain resulting from this decomposition is related, among other factors, to the order of the system and the degree and nature of coupling among the scalar elements. This last aspect influences the rate of convergence, and therefore the number of iterations, called by the supremal, in the decomposed solution.

3.1 _Single Processor Case_

To develop a model for the time gain in this case symbols will be introduced to represent

---

[1] This the practical interpretation of the redefined Hamiltonian.

[2] This involves satisfying the interconnection constraints $\hat{\pi}(\beta) = g(x^*, u^*)$ and as such must yield a higher cost than a similar unconstrained problem.

the extra calculations involved above the classical (not decomposed) solution. These are mainly concerned with the performance measure, the update of the co-ordination variables and the number of iterations needed to produce a solution. These symbols are listed as follows (PT is Processing Time):

τ_i = PT to completely integrate the equations of subsystem i between t_o and t_f

τ_{J_i} = PT to integrate a separate performance measure for each subsystem and include the new term involving β, π and g. There are now N individual measures.

n_i^a = number of iterations required to solve the ith subsystem TPBVP; n_i is the state vector size, and a is a small constant (2 to 3) whose value depends on the optimization process used.

$\tau_{J_{\pi,i}}$ = PT to integrate the ith portion of the supremal error function required to satisfy the model co-ordination variables π - sometimes called J_s. This needs to be done once only, by each infimal after solving its TPBVP.

$\tau_{\pi_{up}}$ = PT taken by the supremal to sum up the error function and update π

I_π = number of iterations called by the supremal to satisfy model co-ordination π

$\tau_{\beta_{up}}$ = PT taken by the supremal to sum up the performance measure $J(\beta)$ and update β

I_β = number of iterations in the outer optimization process needed to maximise $J(\beta)$.

Distinction must be made between the 3 terms "subsystem", "infimal" and "processor". The first will be used to refer to a part of the model or the plant; the second is a theoretical entity responsible for solving the TPBVP of a given subsystem; it may reflect the partial effort of a physical processor, or the total effort of one or more processors.

The total processing time using a single processor is therefore:

$$T_s = \{\{\sum_{i=1}^{N}((\tau_i+\tau_{J_i}) \cdot n_i^a + \tau_{J_{\pi,i}}) + \tau_{\pi_{up}}\} I_\pi + \tau_{\beta_{up}}\} I_\beta \quad (14)$$

Although in general the update terms are small compared with the computation term, they will not be removed from this expression. They will serve as a useful reminder of the nested or hierarchical nature of the algorithm.

(i) Comparison with the undecomposed case.

Let $T = \tau n^a$ be the processing time for the original problem. As an example to get an approximate comparitive figure assume that $\tau_i >> \tau_{J_i}$ (see footnote) and that $\tau_{\pi_{up}}$ and $\tau_{\beta_{up}}$ can be ignored; and that the system is divided into N equal subsystems ($n_1 = n_2..n_N$; and $\tau_1 = \tau_2..\tau_N$); therefore

$$n_1 = \frac{n}{N} \quad ; \quad \tau_1 = \frac{\tau}{N}$$

Substituting in equation 14,

$$T_s = N(\frac{\tau}{N}) \cdot (\frac{n}{N})^a \cdot I_{\pi_{up}} \cdot I_{\beta_{up}}$$

$$= T \cdot (I_{\pi_{up}} \cdot I_{\beta_{up}} / N^a) \quad (15)$$

Therefore a time gain is produced provided the total number of iterations $\ngtr N^a$; e.g. for 10 subsystems (and a = 2) the number of iterations should be < 100.

3.2 Multiprocessor Data Base.

Further time gain can be produced by computing the algorithm on a multiprocessor system. Access to the predicted trajectories of the π and β by the infimal processors can take one of two forms: (i) the collection of all π and β trajectories can be stored in a common memory area accessible by all the infimals through a suitable multiplexing/random-access hardware, or (ii) the subset of π and β used by each infimal is supplied separately by the supremal and stored in the infimal local memory before the start of each iteration.

(i) Common Copy Method. During the computation of the RHS and the performance measure expression, the infimal processors make repeated access to the common copy of π and β. The number and frequency of these accesses are related to the number of times these variables appear in these expressions, and the pattern of their appearance, e.g. the expression for infimal 4 may require 5 successive accesses, while that for 6 may require two accessing periods of 3 and 4 successive accesses separated by a local processing activity.

Using a single common highway to perform these data transfers will ultimately result in congestion and an inevitable waiting time as the number of processors is increased. Let the effect of this waiting time be represented by a finite parameter ω_1 for every extra processor added to the system, so that the iteration time for the ith infimal is now

[3]This is by no means the general case, the reverse is often true. Perhaps to get a more realistic figure τ_i should be doubled which would then halve the maximum number of iterations; down to 50 in the above example.

$(\tau_i+\tau_{J_i})(1+\omega_1 P)$, where P is the number of processors in the system. Similarly, let the effect on the processing time of the model co-ordination error be expressed in terms of a parameter ω_2 so that this time becomes $\tau_{J_{\pi,i}}(1+\omega_2 P)$. The times for updating the common tables of π and β remain unchanged. The multiprocessor time for the common copy method T_{mc} is therefore,

$$T_{mc} = \{\{\frac{1}{P}\cdot\sum_{i=1}^{N}((\tau_i+\tau_{J_i})(1+\omega_1 P)\cdot n_i^a$$
$$+ \tau_{J_{\pi,i}}(1+\omega_2 P)) + \tau_{\pi_{up}}\}\cdot I_\pi + \tau_{\beta_{up}}\}I_\beta$$

(16)

It is noted that in comparison with the single processor equation the computational terms τ_i and $\tau_{J_{\pi,i}}$ have increased by the waiting time factor; but the total sum is decreased by a factor P.

(ii) <u>Multiple Copies Method</u>. The waiting time during computation can be eliminated by supp-lying each processor with its own copy of the subset of π and β trajectories. This, however, leads to the addition of a new step in the algorithm: updating all these copies in the processors local memories. The total update time now consists of two components:
(i) compute the new trajectories, and (ii) store them (multiple copies) in the process-ors' local memories. Let the suffices c and s denote these two components ; so that τ_{π_c} and $\tau_{\pi_{s,p}}$ are the computation time for all the new π trajectories and the storage time into the pth processor respectively. Similarly, τ_{β_c} and $\tau_{\beta_{s,p}}$ stand for the times of the β trajectories. The total multiprocessor time for the multiple copy method is therefore

$$T_{mm} = \{\{\frac{1}{P}\cdot\sum_{i=1}^{N}((\tau_i+\tau_{J_i})\cdot n_i^a + \tau_{J_{\pi,i}})$$
$$+ \tau_{\pi_c} + \sum_{p=1}^{P}\tau_{\pi_{s,p}}\}I_\pi + \tau_{\beta_c} + \sum_{p=1}^{P}\tau_{\beta_{s,p}}\}I_\beta$$

(17)

It is noted that the basic computation time is reduced by a simple factor P (number of processors) compared with the single processor case. The computation times for π and β, τ_{π_c} and τ_{β_c}, may be slightly different from the 'update' times used previously, $\tau_{\pi_{up}}$ and $\tau_{\beta_{up}}$. This depends on the exact code used and processor architecture; e.g. the storage can be an integral part of the last computation step if it involves incrementing the old value by adding the accumulator content to the memory location. The storage time com-ponents in the above expression are summed up over all processors P. Note that there may be more than one 'infimal' (subsystem) alloc ated to a single processor. The subset of π and β trajectories copied into the local memory of a processor need not be the sum total of all those of the individual infimal's assigned to that processor - common trajec-tories are obviously not duplicated within the same processor.

In both of the above two cases, common and multiple copies, it is observed that as a function of the number of processors the total computation time can be viewed as having two components. The first is a decreasing function of P and represents the pure pro-cessing phase; the second is an increasing function of P and reflects the penalty of multiprocessing. This is the waiting time in the first (common copy) case; and update time of extra copies in the second (multiple copy) case.

3.3 <u>Multiprocessor workload division</u>. The number of infimals involved in the analysis of Hierarchical System Theory is derived from a different set of practical constraints to those which determine the number of pro-cessors available for the computation. The first is primarily a function of the degree of coupling among the scalar system elements. These reflect the physical connections among the parts of the real plant and guide the analyst in locating subsystem boundaries in the overall model. Factors influencing the number of processors depend on the mode of operation. In non-real/time (N-R/T) modes the multiprocessor is part of a large comput-ing system available to the design team as a computing aid; and the number of processors available is a function of the instantaneous system workload and may vary with the number of users and the nature of their programs. In real-time systems the number of process-ors involved is a function of physical dis-tances between subsystems, the size of sub-systems, the complexity of their models, and the speed of their dynamics. The computation of the HST algorithm in this case is distri-buted among the processors and computed as a Background job. This is carried out within the remaining portion of the time frame, after executing the real-time control algorithm in Foreground. In both modes of operation the ratio between the number of infimals and processors yields three methods of workload division: (1) more than one infimal per processor; the size of subsystem model may not be large enough to justify a complete processor, or that there is only a limited number of processors which have to be shared out among a large number of subsystems;
(2) one infimal per processor; the sizes of subsystems are roughly the same and there are sufficient numbers of processors to share out,

or that it is easier from a programming point of view to allocate in a one-to-one ratio; (3) more than one processor per infimal: some subsystems are large and require the processing power of more than one processor to solve the problem within specified time limits.

(i) <u>Scheduling</u>. The amount of workload allocated to each processor is related to the number of processors and the time available. In batch-mode operations, the scheduler may assign all processors to the HST program, and the workload is divided evenly among them (as indicated by the 1/p term in the two expressions of the last section). Of course it is highly unlikely that all processors can be given the same work load (subsystems are not usually similar and their program execution time may vary between iterations). In on-line/interactive applications the scheduler allocates a number of processors to the algorithm for a specific period of time. The algorithm may not have fully executed at the end of this period, with the inevitable situation of having some or all processors withdrawn. At some later time instant the scheduler may be able to restore, or increase, the number of processors allocated to the algorithms. The number of infimals per processor is therefore time varying and affects the total algorithm solution time.

In real-time/computer-control environments the situation is slightly different. Each processor here has a pre-assigned work load which must be executed repeatedly at a pre-defined clock-rate. These work loads/clock rates (foreground jobs) combinations may be different for each processor, and result in different size portions of frame times available for the execution of the HST algorithm. Furthermore, the size of the real-time jobs may vary according to the state of the plant, and thus giving variations in the time left for background work. An efficient scheduler is therefore required to monitor the end of the foreground work in each processor in the system, and assess the distribution of available processing power remaining in the system until the start of the next frame. The HST algorithm, or what remains of it to be computed from previous cycles, is then remapped onto the network according to this distribution. This means that processors with smaller remaining processing power are given correspondingly smaller numbers of infimals, or infimals with simpler models. As interaction among units at the infimal level is absent, adjacency between processor-allocated infimals need not correspond to subsystems neighbourhood. It must be emphasised, however, that this is not so when more than one processor is assigned to an infimal—method 3 above. Interaction at the sub-integration cycle level is required in this case to complete each single evaluation of the subsystem RHS.

(ii) <u>Infimal/Processor Distribution</u>. A systematic design of a scheduler involving the above concepts can be aided by the introduction of the following symbols. These represent the number of π and β trajectories, the number of infimals per processor, and other variables. They are briefly outlined here to clarify the ideas discussed and will not be pursued further in this paper.

S = size of total (π,β) vector, $2(n+m)$

S_i = ith infimal (π,β) vector size, (related to the exact form of $\sum_j \beta'_j g_{ji}$, and not $(n_i + x_i)$)

S_p = number of (π,β) trajectories stored in the pth processor

N_p = number of infimals assigned to the pth processor

i_p = the number of the first infimal handled by the pth processor

τ_p = processing time available from processor p for background jobs

The following simple relations follow:

(a) the last infimal handled by the pth processor is $i_p + N_p - 1 = i_{p+1} - 1$

(b) the total number of infimals N is:
$$N = \sum_{p=1}^{P} N_p$$

(c) the number of π,β trajectories to be accessed by the pth processor is:
$$S_p \leq \sum_{i=i_p}^{i_p + N_p - 1} S_i$$

i.e. common trajectories among the infimals of the same processor are not duplicated.

(d) the number of infimals assigned to pth processors, N_p, must be computed within its available processing time
$$\sum_{i=i_p}^{i_p + N_p + 1} ((\tau_i + \tau_{J_i}) n_i^a + \tau_{J_{\pi,i}}) \leq \tau_p$$

4. DISTRIBUTED COMPUTER CONTROL

Implementing controllers on multiple processors rather than a single large computer may be attributed to three interrelated reasons: functional, economic and technical. The first of these is implied in the underlying theme common to the concepts discussed so far: that the plant by its physical nature is made up of smaller subsystems, where couplings between the state variables of a subsystem are generally stronger than couplings between subsystems.

In practice this also implies relatively greater distances between subsystems, and lead to the viable engineering solution of allocating one processor to each subsystem. The economic and technical reasons play their part here, where large scale integration technology has resulted in cheap and small processing units which may be distributed throughout the plant.

4.1 Design Features. The selection and installation of microcomputers as a network of distributed-controllers is influenced by two major factors: the form and make-up of the plant to be controlled; and available know-how and technology in the processing, memory, interface and communication hardware and software.

(i) Plant. Relevant design parameters include (a) the physical size of the plant; (b) the degree of distribution and the distances involved; (c) the number of subsystems and the decision factors influencing the position of subsystem boundaries, e.g. the number, frequency and degree of electrical/mechanical coupling, distances, etc.; (d) number of inputs/outputs; (e) complexity of model, stationary/time-varying/non-linear/stochastic; (f) speed of response; (g) sampling rate and the need for multiple rates to cope with input/outputs of differing speeds; (h) size and distribution of actuators and sensors; some of these may be of such size and/or complexity as to warrant an individual controller in their own right.

(ii) Controller hardware. The number, size and pattern of distribution of the processors executing the control algorithms is influenced as much by hardware and software limitations as by the characteristics of the system to be controlled. Design aspects involved in the processor hardware include (a) the speed limitations of semiconductor fabrication technology, CMOS, NMOS, $I^2L^{(3)}$, Schotcky, SOS, Bipolar, etc.; (b) the processor word length, 4, 8, 12, 16, 32 bits; (c) processor instruction set: primitive instructions, high level instructions (transcendental functions), combination instructions (decrement, test and branch if zero), multiple operand instructions; (d) microprogrammed/hardwired sub-instruction execution; (e) floating point arithmetic hardware; (f) vector/matrix operations hardware; (g) special purpose handwired logic to execute very fast algorithms; (h) relative speed of processor to memory; (i) size and speed of memory; (j) type of memory, RAM, ROM, PROM, EPROM, core; (k) size and speed of backing store, magnetic disk (hard, floppy, mini), magnetic tape, bubble memory.

(iii) Software. The above factors influence the selection of the individual processors controlling one or adjacent collection of subsystems. The programming, testing and validation of the control algorithms resident in these processors is influenced by another set of factors: (a) availability of real-time languages, assembler, machine code/mnemonic level, medium or high level languages; (b) cross-assemblers/compilers on a host machine accessible to the design team; (c) processor simulators to run and test these programs and remove, as far as possible, any programming errors; (d) plant simulators to run and test the algorithms embedded in these programs and remove errors and/or adjust parameter values to meet specifications; (e) availability of simple operating systems in the controller processors to enable loading and running of these programs and subsequent final tuning of parameters, and, if necessary, re-assembly or recompilation in case of program/algorithm modification.

(iv) Networks. Reasons for interconnecting the processing elements of a distributed real-time computer control system bears only some resemblance to motivations for national and international networks. Factors influencing the present case include: (a) the control algorithm of a particular subsystem requires the output (or input, or both) of one or more other subsystems which are controlled by separate processors; (b) the operation of parts of the plant must be synchonised to ensure a correct transition between the various phases of execution of a given sequence, e.g. start-up, shut-down and predefined fault conditions; the algorithms which deal with these situations and are resident in various processors controlling the different parts of the plant, must exchange the appropriate status and synchronisation signals; (c) the processor may need to be interrogated from a higher level, but distant, controller; it would be cheaper in this case to connect the processors onto a common highway leading to the controller rather than have a separate line from each processor to the controller; (d) the programs developed at a central mainframe for the local processors need to be down-loaded to the local memories; again a common highway is a more cost effective method of performing this transfer; (e) during the test and validation stage of the programs/algorithms an engineer working at the local subsystem processor level may require to modify, re-assemble/recompile,relink and reload small parts of the program; the overall data processing system design may offer one of two alternatives: the engineer's portable terminal may be linked via the local processor and common highway to the central mainframe to give him all the usual computer room facilities; or the local operating system may trigger the central computer to down load system and source files to perform all the subsequent file operations at the local level. There are two points to observe here, (1) bubble memory densities and fibre-optic link bandwidths make the second alternative viable, which has the attraction of independence from central mainframe work load; (2) source file integrity must be preserved by copying modifications onto central file for documentation maintenance; for either alternative the common highway method is again an attractive proposition.

(v) <u>Co-ordinated distribution of background jobs</u>. A loosely coupled real-time multiple processor system may be considered for performing non-real-time tasks concurrently. Motivations for this are many folds: (a) there may be periods in the 24-hour cycle when the plant is relatively quiet and the distributed processing power can then be used for other jobs; (b) even during the plant normal operation there may be spare power (the processors are usually chosen to handle the heaviest forceable work load; furthermore, it is generally found to be cheaper and more convenient (technical and administrative constraints such as trained service personnel and spares) to standardise on a particular powerful processor for all the elements in the network thus leaving plenty of spare power in the system; (c) the plant is often time-varying and may require new trajectories to be calculated to maintain optimum performances; it would be convenient to perform this computation within the same processing system, rather than resorting to the computing power of some remote mainframe; (d) the algorithm of hierarchical system theory demands much less computation than is required for undecomposed systems, so the need for superfast mainframe is alleviated. In this case a collection of relatively slower processors tightly co-ordinated by a 'supremal' seems to be well suited for the purpose.

4.2 <u>Interconnection Schemes</u>. Detailed examination of the aspects involved in this distributed time-sharing of work relies heavily on two interrelated areas. The first entails a full assessment of the data processing needs of the task to be performed in the background. The second area attempts to reconcile these needs with available techniques in programming, in devising special instructions to speed up and simplify co-ordination and in formulating special bus structures.

(i) <u>The data processing requirements of hierarchical systems theory</u>. Identifying these needs involves computational as well as data transfer considerations at many levels. The arguments underlying these needs may be developed by considering the following factors (a) the transfer of data (π, β, $\hat{x}^*$, $\hat{u}^*$) between the supremal and the local memories of the infimals must be fast and efficient to represent as small a portion of the total iteration time as possible; (b) a relatively large subsystem model may not be ameniable for further decomposition and optimization at the infimal level; the model RHS will then have to be partitioned for multiprocessor computation at the sub-integration cycle level. The distinction between this partitioning and decomposition into subsystems must be emphasised. In decomposition the infimal treats the subsystem as an independent and complete system which must be optimized according to given constraints. This generally means solving the system equations (once for each iteration in the optimization process) using a suitable integration routine, and may involve hundreds of integration cycles for each run. Partitioning is aimed at this, integration cycle, level and attempts to reduce the computation time of each evaluation of the RHS. The speed and efficiency of data transfers at this level are far higher than rates associated with distributed networks, and is crucial to the achievement of worthwhile time gains. Furthermore, it is noted that the data transferred is not related to the co-ordination variables (π and β), but are the state variables of the ith infimal $\underline{x}_i$ which generally require stronger coupling among the associated scalar elements than is required for model and goal co-ordination variables. (c) An arbitrary number of processors may be required to co-operate in this fashion to aid a given infimal; (d) there may be several infimals whose individual work loads require assistance of several processors; (e) the data transfers among the processors of a given infimal must carry on uniterrupted, and at the highest possible rate; thus independence between infimals is necessary to ensure transfers in parallel; (f) if a common highway, or bus, is to be used then a mechanism must be provided for isolating sections of it temporarily to carry out parallel transfers; (g) this isolation must be under the control of the processors in the network, and fast enough to enable efficient switching at the infimal/supremal iteration level; (h) no one processor, or collection of processors, in the network can be pre-assigned to be a main infimal processor as the number of infimals, and processors per infimal, is problem dependent and thus arbitrary.

(ii) <u>A one-dimensional structure</u>. The requirement of multiple/parallel data transfers in arbitrary size clusters can be neatly realised by associating a common bus switch with each processor in the network, see AlDabass, 1978. The control of these switches however, presents a formidable problem. For example, if each switch was placed under the control of its own processor, then it would need a considerable amount of message passing and synchronisation operations among the processors to achieve a given cluster formation. Another problem is that of cross transferring data among the processors of a cluster in a fast and efficient way. Furthermore, these data transfers must in no way create deadlock and/or conflict situations which can only be resolved through the intervention of operating system staff. In essence a fundamental and unifying concept is looked for which would serve both as an interconnection structure and as a philosophy to organize, map and synchronize the workload onto the multiprocessor system.

Hierarchy is one such concept, which has proved, through extensive design and simulation exercises (AlDabass, 1977a,b,1978b) to be easily applicable to multiprocessor structures. To eliminate conflict and deadlock completely, only one unit is placed in each level of the hierarchy. A processor at any level can operate in one of two modes: master, in complete control of all the remaining units below it in the system; or slave, ready to receive orders from any unit above it in the hierarchy. These orders include the control of common bus switches, and signals to connect the memory of any unit directly to the common bus for the cross transfer of data. Special microprogrammed instructions are developed to ease the programmer's task of synchronising processes resident in the different processors, and to access data efficiently.

A typical sequence for loading and executing a program in a non-time-sharing mode is as follows: (a) the overall master closes all switches and connects all memory units to the common bus; (b) programs and data (object files) are loaded from a mass storage device into the local memories; these programs also include sections which synchronise the processes; (c) the master starts all processors and opens all switches; (d) some processors are programmed to halt upon completing the execution of their processes ; others are programmed to monitor the halt state of these and access their results before entering a second phase of processing; (e) the overall master monitors the halt state of all other processors, closes all switches and connects all local memories to the common bus to access the final results; the system is then ready to start another cycle.

(iii) <u>Time-sharing</u>. The execution of the algorithm of Hierarchical System Theory can be very easily tailored to fit the above procedure. The supremal role can be programmed on a unit nearer the upper regions of the hierarchy, while the infimals are programmed on lower units. The supply of predicted trajectories from the supremal to the infimals is performed by closing all switches and connecting the infimals' memories to the common bus. The infimal processors are then started and all switches opened. Infimals consisting of more than one processor are mapped on adjacent processors, with a local master being programmed with the infimal optimization process, while the subsystem model and performance measure are partitioned and distributed within the memories of the assisting processors to form a local cluster.

To time-share this procedure with a real-time job in the foreground is straightforward in concept, although it may need additional hardware. The idea is to keep the optimization process running at a low priority and assign a higher priority to the clock interrupt which triggers the real-time control algorithm. On the completion of the execution of this algorithm, a normal return-from-interrupt is performed which restores processor control back to the HST optimization process.

Additional hardware is required to prevent interruption of the real-time task by units higher up in the hierarchy; i.e. the relevant processor must be removed from the hierarchy for the time period when the foreground job is being executed. In general the common bus section associated with the processor remains free during this period and may be used by other units to cross-transfer data. However, to ease scheduling and prevent the time wastage of waiting for a processor to complete its real-time work, it may be easier to synchronise the start of all foreground jobs to a master clock. The idea being that all the spare processing power is left to the later portions of the time frames, thus minimising the likelihood of a background data transfer being delayed by a real-time processor which is temporarily absent from the hierarchy.

A major background task that has to be performed before any other non-real-time work, is the update of input/output from other parts of the plant. This data is used in the real-time algorithm and must be updated before the next pass of this algorithm begins. The processor which happens to be higher (in the case of two processors) in the hierarchy performs this transfer by waiting for the other to finish its real-time work. On completion of this transfer, the processor starts the background work proper - the HST algorithm in this case.

5. CONCLUSIONS

The feasibility of computing the algorithm of hierarchical system theory as a background job on a real-time computer control network was considered in some detail. Of the three methods proposed for allocating infimals to processors, the 'multiprocessor-infimal' method proved to impose the severest requirement on the interconnection structure. A dynamically segmentable common-bus scheme was put forward to meet these needs. A continuous hierarchy imposed on the elements of the multiprocessor network was suggested, both to eliminate deadlock and conflict, and as a concept to aid the programmer in mapping out the background work load onto the network. Considerable work remains to verify these concepts through detailed design studies and simulation; and to extend and modify them by examining the influence of other areas in control, such as large scale state estimation, identification and adaptive control policies.

ACKNOWLEDGEMENTS

The author wishes to thank the Science Research Council for the fellowship to carry out this work. Thanks are also due to Professors Munro, Singh, Aspinal and Kropholler, and other colleagues for stimulating discussions and encouragement.

REFERENCES

Adams, G. and T. Rolander (1978). Design motivations for multiple processor microcomputer systems. Computer Design, Vol. 17, No. 3, pp. 81-89.

AlDabass, D. (1976). Two methods for the solution of the dynamic programming algorithm on a multiprocessor cluster. Report No. 347, Control Systems Centre, UMIST, Manchester.

AlDabass, D. (1977). Microprocessor based parallel computers and their application to the solution of control algorithms. ACM International Computing Symposium, April, 1977, Leige, Belgium, pp.261-270.

AlDabass, D. and D. Rutherford (1978). Simulation of a chain-like multi processor for continuous systems simulation. Proc. of UK Simulation Council Conference on Computer Simulation, Chester, April, 1978, pp. 338-349.

AlDabass, D. (1978). A Reconfigurable Parallel Processing Structure for adaptive state estimation. Control Systems Centre Report No. 415, UMIST, May 1978.

AlDabass, D. (1979). Simulation of a two-dimensional multiprocessor structure. Control Systems Centre Report No. 446, January, 1979.

Amrehn, H. (1974). Digital computer applications in chemical and oil industries. 4th IFAC/IFIP International Conf. on Digital Computer Applications to Process Control, Pt. III, pp.176-194. (Ed. M. Mansour & W. Schaufelberger).

Baer, J.L. (1974). Models for the design, simulation and performance of distributed function architecture. IEEE Computer Workshop on Distributed Functions Arch.

Bauman, E.J. (1968). Multilevel optimization techniques with applications to trajectory decomposition. Advances in Control Systems (Ed. C.T. Leondes), Vol. 6, Academic Press.

Bell, D.J. and D.H. Jacobson (1975). Singular optimal control problems. Academic Press.

Coleman, M.L. (1973). ACCNET - a corporate computer network. AFIPS National Computer Conference, 4-8 June 1973, New York, pp. 133-140.

Down, P.J. and F.E. Taylor (1976). Why distributed computing. NCC Publications.

Fabian, M. (1978). Communicating with pulsed lasers. New Electronics, Vol. 11, No.14, pp.26-33.

Gertler, J. (1975). The practical apparatus of process control. IFAC 6th Triennial World Congress, Boston, USA, 24-30 Aug., 1975, part IVB (Computers, Space).

Gill, P.E. and W. Murray (Eds.) (1974). Numerical methods for constrained optimization. Academic Press.

Goguen, J.A. (1970). Mathematical Representation of hierarchically organized systems. In Global systems dynamics, edited by E. Attinger and S. Karger, Basel, Switzerland, pp. 112-129.

Hayes, J.P. and R. Yanney (1978). Fault recovery in multiprocessor networks. Session 8 : Distributed systems, 8th International Symposium on Fault Tolerant Computing, (IEEE), Toulouse, France, 21-23 June, 1978.

Hnatek, E.R. A User's Handbook of Semiconductor Memories. Wiley-Interscience.

IEE (1978). Software Engineering for Telecom'n switching systems. Sessions 8,10, 11 on Operating Systems Compiler Techniques and Distributed Control. Conference Proc. of 3rd International Conf., Helsinki, Finland, 27-29 June, 1978.

IEEE Press Selected Reprint Series (1976). R. Blanc & I. Cotton (Eds.), Computer Networking.

Jacobson, D.H. On computational methods for dynamic and static optimization. AFIPS 1972 SJCC, vol. 40, pp.181-185.

Jervis, M.W. (1972). On-line computers in power stations. Proc. IEE, IEE Reviews, Vol. 119, No. 8R, pp. 1052-1076.

Jones, P.D. (1977). Criteria for selecting and interconnecting micro, mini, midi and maxi computers. ACTA Cryst., A33, pp.20-24.

Kriloff, H.Z. (1973). A high level language for use with multi-computer networks. AFIPS Proc. National Comp.Conf., New York, June 4-8, 1973, pp. 149-153.

Laniotis, D.G. (1978). Partitioned controls: fast, parallel processing algorithms. AFCET/IFAC Symp. on systems optimization and analysis, France, 11-13 Dec. 1978.

Lauer, P.E. (1977). Abstract specification of resource assessing disciplines and management strategies. Workshop on Global description methods for synchronization in real-time applications, organized by the Real-time Systems Group, Information Div., AFCET, Paris, 3-4 Nov. 1977.

Luenberger, D.G. (1973). Introduction to linear and nonlinear programming. Addison-Wesley, Massachusetts.

Macko, D. (1966). A co-ordination technique for interacting dynamic system. Proc. JACC, pp. 55-60

Mehra, R.K. and R.E. Davis (1971). A generalised gradient method for optimal control problems with inequality constraints and singular arc. Proc. JACC, pp. 144-151.

Mesarovic, M.D., D. Macko, and Y. Takahara (1970). Theory of hierarchical multilevel systems. Academic press.

Murray, W. (Ed.) (1972). Numerical methods for unconstrained optimization. Academic Press.

Nicholson, H. (1978). Structure of interconnected systems. Published by Peter Peregrinus Ltd., England.

Ramamoorthy, C.V., K.M. Cahndy, and M.J. Gonzalez. (1972). Optimal scheduling strategies in multiprocessor systems. IEEE Trans. on Comp., pp. 137-146.

Roberts, L.G. and B.D. Wessler (1973). The ARPA Network. In Computer communication networks, Prentice-Hall, pp.485-499.

Sage, A.P. (1968). Optimum Systems control. Prentice-Hall.

Schwartz, M. (1978). Computer-communication network design and analysis. Prentice-Hall.

Singh, M. (1977). Dynamical hierarchical control. North Holland.

Singh, M. and A. Titli (1978). Systems, decomposition, optimization and control. Pergamon Press.

Smith, N.J. and A.P. Sage (1973). An introduction to hierarchical systems theory. International Journal of Computers & Electrical Engineering, Vol. 1, No.1, pp. 55-72.

Steusloff, H. (1976). Hardware structures and software for distributed computer control systems. Proc. 3rd Annual Advanced Control Conf., Purdue University, pp. 19-48.

Stokes, A.V., D.L. Bates, and P.T. Kirstein. Monitoring and access control of the London Node of ARPANET. NCC, pp.597-603.

Takahara, Y. and M.D. Mesarovic (1969). Co-ordinability of dynamic systems. Proc. JACC, pp. 877-889.

Thomesse, J.P. (1978). A new set of software tools for designing and realizing distributed systems in process control, vol. 8. Annual review in Automatic programming, Ed. M.I. Halpern.

Vichnevetsky, R. (1978). On the difficulty of computing optimal control problems. Proc. of IMACS Symp. Simulation of Control Systems, Vienna, 27-29 Sept.,1978, Ed. I. Troch.

Wismer, D.A. (Ed.) (1971). Optimization methods for large scale systems. McGraw-Hill.

DISCUSSION

Buchner: My first question is on the control problem. You have a particular distributed control system and you noted that there were some spare time slots for some computation. I was wondering if the control that you are implementing with your control tasks, C1, C2 or C3 and so, is the same process for which you are trying to develop control by solving the particular optimal control problem?

AlDabass: Yes, it is. One interpretation of using optimal control theory is that we are trying to generate the optimal trajectory for a subsystem. This calculation is done in the background, and once we have it, we use this trajectory as a changing reference point for the subsystem controller. Thus, the subsystem output will follow the optimal trajectory that we have computed.

Buchner: Given that, the solution of the optimal control problem provides an open loop solution, and not a feedback solution. So why can this not be done prior to implementing the control, or are you considering some type of adaptive approach?

AlDabass: Oh yes, this is the whole idea really; the plant itself is not fixed. It could be time varying in such a way that we can't precompute this optimal trajectory. We have to compute it regularly, say once a day or whenever. In fact, we monitor changes in the parameters of the plant or the system which would warrant recomputation.

Buchner: A major point behind your discussion of the solution of the differential equations is the requirement for maintaining a data base consisting of the state of the process, such that all the states are represented at the same exact time. You spoke about finishing all the individual processes before you go on and make another update.

I'm wondering if that is really all that important, because in numerically solving differential equations through integration techniques, one can take the approach that one is not particularly concerned about the state variables at a single instant and that there are tradeoffs between how much you are concerned about maintaining synchronism and the amount of time that you can let the processes proceed asynchronously. One might allow larger integration intervals and have an increase in the error, as well as larger amounts of asynchronism with corresponding increases in your error. You really have both things to play with and it may be an important avenue of research to investigate both possibilities.

AlDabass: Yes. The problem is that when you are solving vector differential equations, you really must maintain consistency. Let's say, for example, that we have a simple differential equation involving the variable $x1$ and its derivative $\dot{x}1$. Now $\dot{x}1$ is usually a function of, not only $x1$, but of the other state variables in the system, let's say $x2$, $x3$ and so on. We really cannot maintain the integrity of the solutions if we do not update the values of $x1$, $x2$, and $x3$ in finding the value of $\dot{x}1$. There could well be some mathematical mapping of some kind, whereby you could end up with a correct solution, even if you do lose synchronism between the values of $x1$ and the values of the components.

Buchner: I'm just saying there is nothing magical about using a variable at the previous time period and maintaining it. There are possibilities to do this and I've done some of that work by accident in improperly structuring integration routines. I've gotten very reasonable results, so it indicates some possibilities.

AlDabass: Well, as a possible research area, it certainly is a valid one.

Humphrey: In your discussion, you showed an optimization curve of the number of processors versus a performance parameter. You imply that the curve is a parabola. Do you really believe that it is a parabola? That it is going like some number to the minus n squared power?

AlDabass: No, it is not meant to be a parabola specifically. It is just meant to show that there is a minimum.

Humphrey: Do you have any idea what behavior you would expect from your particular model?

AlDabass: Yes, it would show the usual minimum; but the exact values at the minimum are functions of the parameters in the model.

Humphrey: The thing is that this is the first model that I have personally seen in which it seems to me that you've got a real chance for proving that there is an optimum number of processors for some specific distributed processing model. I would just like to have some sort of feeling about the kind of behavior that goes with the number of processors.

AlDabass: I've done extensive computations for a particular problem; namely, optimization using Bellman's dynamic programming. I've formulated a model for a multiprocessor structure in that case and they all look like parabolas.

They show a sharp decrease for a small number of processors, and then start to increase somewhat less sharply beyond the minimum. Basically this is what I tried to depict but, of course, it all depends on the exact nature of the overhead and other factors.

Sheane: Is there an opportunity to organize the processors in such a way that an individual processor may fail and the surviving processors will continue with the required computing capacity?

AlDabass: This is really a reliability issue you are talking about, versus pure performance increasing procedures. So far I've only looked at the performance part of it, but I'm hoping to look at the reliability aspect in the future and see how the interconnection structure that I've put forward stands up to that.

Willard: In your model of the hierarchical system, I believe you implied that elements in the middle of the hierarchy must be either in master mode or in slave mode. Yet, the model of a hierarchical system that I believe most of us carry in our heads is that middle level elements must concurrently be master of elements below then and slave to at least one element above them. Could you clarify your position on that point?

AlDabass: I think the key word in your question is "concurrency." There is a lot of debate on what concurrency means, whether it really means simultaneity. For example, if you look within a particular sampling period of] millisecond then, within that] millisecond period, you could alternate between the slave mode and the master mode. But, if you are looking at the system only at the one millisecond points, then, for all intents and purposes, that system has operated in the slave and master mode concurrently. But, when you go down to the more detailed level, you find that it really hasn't; there is a specific part of the real time where it was exclusively in a master mode and then there was another portion of real time where it was exclusively in a slave mode.

Gellie: There is an implication that the topology and the number of processors that you might use in a DCCS could be in some way governed by the work you have been talking about if one wanted to have continuous optimal control calculations. If so, one would have to relook at the network and topology, as originally conceived as being ideal by the control engineer. This sounds like we've bridged the gap between DDP and DCCS.

AlDabass: Yes, perhaps there are implications there. I suppose there are two ways of looking at it: If we look at it from the background angle, where we are computing a job in the background, and we find that there is an optimum number of processors for the computation of that task, then we can make the number of computers in the DCC network the same as the optimum number. Or, we could also say that there is no need to utilize all the processors in the DCC network for that. If there are more than the optimum number, then we just concentrate on using the optimum number and leave all the rest.

Johnson: I have one comment and one question. The comment follows up on the remark of Mark Buchner.

We can make a closer tie, I think, between the offline computation that is normally associated with the pure optimization problem and an online computation which most of us have been discussing in the computer control networks. One way of implementing an adaptive controller is to try to make these computations recursive in time; at the same time that you are doing the optimization you are also doing recursion. Now if we consider the situation where we are trying to tune, say, a local PID loop using its own processor at the same time that the setpoint of that loop is being generated by an overall optimization program at the plant level, which is analogous to your determination of Pi and Beta, we get a structure which is probably more familiar to people in working in process control.

Now the question. It seems that you have implicitly assumed a communications system which is parallel, in the sense that the master slave information can be passed along some control lines at the same time that the data is being transmitted over a bus. In other words, you haven't allowed for the possibility of a serial communications protocol, have you?

AlDabass: This is quite so. I haven't. The origin of this structure stems from my early work on parallel processors. What I am trying to do is to extend the structure that was developed for clustered parallel processors (that is, processors all sitting within the same cabinet) to a distributed network case and see how it works and how we can use it. Having said that, I can't see anything inherent in the way the hierarchy idea works that stops it from being used with a serial communication protocol.

THE DESIGN OF DISTRIBUTED COMPUTER CONTROL SYSTEMS

A. Nader

Department of Computing and Control, Imperial College of Science and Technology, London, U.K.

Abstract. The complexity of decentralised control, large scale systems and distributed processing are combined in the design of distributed computer control systems. A general view of the problems that arise in the design of distributed computer control systems and some of the present approaches to solve them are presented. Particular emphasis in the cross fertilization of computer science and control theory is made.

Keywords. Computer Control; Control engineering computer applications; Decentralised Control; Distributed Computer Control systems; Hierarchical systems; Industrial Control; Multiprocessing systems; Parallel Processing; Process control.

INTRODUCTION

The design of distributed computer control systems (D.C.C.S.) involves relatively new areas in control theory and computing science: Decentralised Control, Large Scale Systems, Real Time Distributed Systems.

Because D.C.C.S. have only recently become feasible, the above areas are far from being completely developed and understood. The current theoretical understanding of D.C.C.S. reflects the state of centralised control and computer theories. In many cases the theory behind centralised systems is not fully developed. For instance in control theory, centralised linear control systems are well understood, but the theory of nonlinear control systems is not well developed. This is clearly reflected in decentralised control. In computing science the theory behind languages and operating systems for real time centralised computer systems is not fully developed and this has reflected in the lack of understanding of operating systems and programming languages for distributed systems.

This paper reviews some design problems posed by a D.C.C.S., the state of the art of the present solutions and methodologies to solve them and, most important, the research directions in the use of tools and methodologies from control theory in computing science and vice versa.

This review is not intended to be exhaustive; it is intended to give a general picture within a particular design framework.

It also intends to introduce some of the "unconventional" approaches to solve design problems in D.C.C.S. Recent research has shown that the tools and methodologies used in both computing and control could be complementary to each other in order to solve some of the D.C.C.S. design problems.

It intends to display a conceptual unity among computer and control design problems structured around a particular design approach.

The Design of D.C.C.S. Framework

Real time process control systems were designed in the past around general purpose real time computers (which are a compromise between various supposed needs). The software system was adapted to a class of problems. The designer was left the creation of the particular application.

The opportunity to build specific computing units according to specific needs has made a change of design strategy. Today's design approach is more problem orientated. After the definition and design of a particular application there is an implementation choice for it (hardware/software). A global top down approach in the design of the D.C.C.S. is now feasible.

D.C.C.S. Definition (Nader, 1979a)

Let M be a set $M = \{M_1, \ldots M_m\}$ of modules that perform operations necessary to achieve the control of the plant. The control of the plant being specified by a set of control functions $F = \{f_1,...f_f\}$.

We define a D.C.C.S. as the control system whose set of modules M is distributed among a set of computation location $C = \{C_1,...C_c\}$.

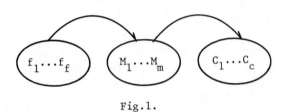

Fig.1.

Some of the classes of control functions can be:

- computation of the control actions
- sequencing
- start up
- shut down
- exceptions procedures
- supervisory control
- monitoring
- information display etc.

Note that the control functions have a more general meaning than computation of control actions. D.C.C.S. are defined on implementation grounds.

To achieve the control of the plant with a D.C.C.S. some information communications take place. We characterise the control functions by their "information pattern". The information pattern is constituted by:

- implicit information (design decisions)
- structural information (non changing information while the system is in operation)
- functional information (changing information while the system is in operation)

$IP = \{II, SI, FI\}$.

The information pattern is the most important characteristic of D.C.C.S.

D.C.C.S. Design Procedure

The different design steps of the D.C.C.S. are presented in fig.2.

Note that decomposition can occur at different steps in the design procedure.

In the first section of this paper, we present some aspects of the design of the distributed control system:

Modelling of the system

    model simplification, reduction
    model evaluation,
    model decomposition

System Analysis

    stability, structural stability,
    controllability, observability

The Distributed Control System

    decentralised observation and estimation
    stabilisation and pole assignment
    decentralised optimal control
    multilayer and hierarchical control
    game theoretic approaches in optimal control
    centralised control algorithm decomposition

In the second section of this paper we present some aspects of the design of the distributed computer system:

Distributed Processing

Concurrent Programming

    The layered structure
    The module concept
    Interprocess communication and Synchronization
    Protection and correctness

Distributed Real time Operating System

    Scheduling

Application software

    software decomposition
    requirements specification
    reliability and correctness

Communication Networks

I

THE CONTROL SYSTEM

MODELLING

The first step in the design of a distributed computer control system involves the obtention of suitable models of the plant for analysis and design of the control system. These models can be obtained either by modelling through established laws (chemical, physical, etc) or by identification of the plant.

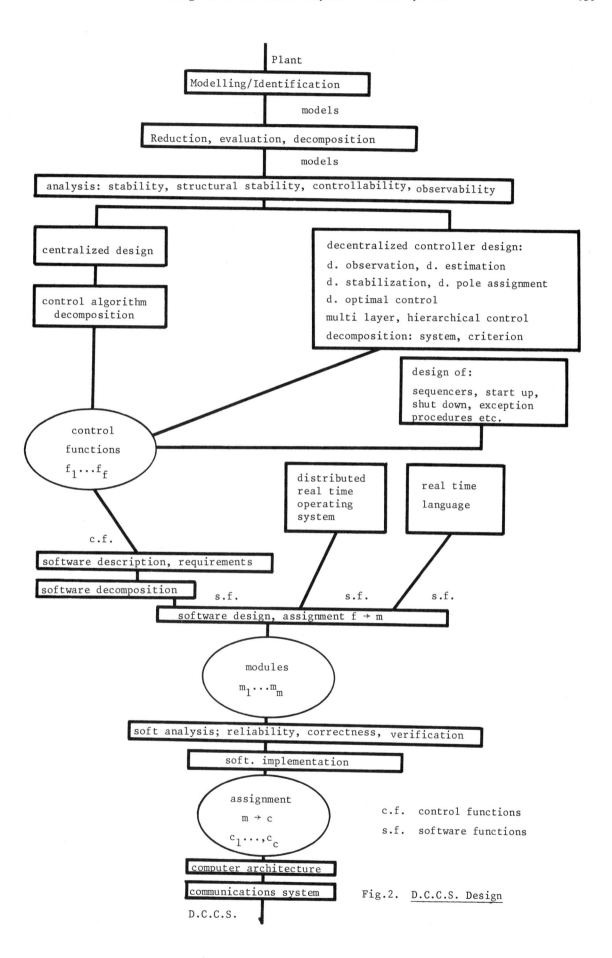

Fig.2. D.C.C.S. Design

The plant could be perceived by the designer either

- as a set of interrelated subsystems. This is the case where most functional, geographical or topological decompositions take place.
- as a single system.

Essential operations necessary to design a D.C.C.S. at the modelling stage are:

- Model reduction
- Model evaluation
- Model decomposition

Model Reduction

As with other studies in D.C.C.S. this operation has been studied from two different points of view: as an additional (extension) issue in modern control theory, and from less orthodox approaches such as information theory and complexity theory.

In modern control theory, model reduction is restricted to the linear case.

The problem of model reduction or simplification is:

Let M be a linear irreducible in order n model. The problem is to find another model $\bar{M}$ of order $m < n$ that approximates model M.

Different criteria can be used for the obtention of model $\bar{M}$:

- conservation of modes
- steady state value (static error, properties)
- difference $y(t) - \bar{y}(t)$ (minimization criteria)
- structure preservation
- stability properties
- quality insensitive to a class of inputs
- calculations involved

The methods that can be used to achieve model simplification can be of the type

- Pade approximations
- optimal reduction min $j = \int_o^\infty (y-\bar{y})^2 dt$
- state vector approximation
- aggregation methods
- perturbation methods

Model Evaluation

Another less orthodox approach in model reduction is: given several models of the same plant evaluate them in order to choose the simplest. The criteria used in this evaluation could be:

- entropy from information theory
- complexity (algorithmic)
    - Kolmogorov
    - cyclomatic

Kolmogorov complexity and entropy

In (Petrov and colleague, 1978) the physical complexity of a dynamic system is measured as the number of independent parameters on which the system entropy depends.

The Complexity $K(x)$ of a finite plant x is represented as the minimal number of bits (Minimal program length $l(p)$) which contain all data on the plant sufficient for its restoration.

The Complexity $K(x)$ intuitively defines the amount of information necessary to restore the plant x.

The Kolmogorov complexity $K(x)$ and the Shanon informational entropy are related by:

$$K(x) \leq i \, (H(qk) + \alpha(i))$$
$$H(qk) = \sum_{k=2}^{2^r} \log_2 qk$$
$$\alpha(i) = Cr \, \frac{\ln i}{i} \text{ as } i \to \infty$$

Another similar approach is presented in (Maciejowski, 1978) where a method of selecting between competing models of small sets of observations is developed. A model defined as a program allows one to measure the size of a model in terms of the size of a program and then interpreted as a measure of the informativeness of the model. The Kolmogorov complexity and the Shanon informational entropy are also used. The approximation - complexity trade off is also studied.

Cyclomatic Complexity

In (McCabe, 1976) the authors provide a quantitative technique for modularization based on program flow control.

The complexity measure approach taken is to measure and control the number of paths through a program.

The Cyclomatic number $V(G)$ of a graph G with n vertices, e edges and p connected components is

$$V(G) = e - n + p$$

In a strongly connected graph G the maximum number of linearly independent circuits is the cyclomatic number. In order to get for any program a strongly connected graph an external edge between the beginning and the ending of the program is added.

The cyclomatic complexity is defined as

$$V(G) = e - n + 2p$$

The cyclomatic complexity measure can be used as the Kilmogorov complexity to choose competing models.

Furthermore the complexity ideas can also be used in model decomposition.

Model Decomposition

Different methodologies are used in order to decompose models of dynamical systems:

- graph theory approaches
- information theory approaches - Kolmogorov complexity
- category theory approaches
- automata theory approaches

Some graph theory approaches

In (Caines and Printis, 1978) the authors considered a special class of variable parameter system given by the equations:

$$\frac{dXs}{dt} = (fs^0(xr_1,\ldots xr_v) + As)Xs$$
$$+ fs^1(xr_1,\ldots xr_i, xr_v) + Bs)Us$$
$$ys = (fs^2(xr_1,\ldots xr_i, xr_v)X + Cs)Xs$$
$$+ fs^3(xr_1,\ldots xr_i \, mxr_v) + Ds)Us$$

where $xr_i \varepsilon \{xr_1 \ldots xr_v\}$

if $r_i < S \quad 1 \leqslant i \leqslant v$ and, (Cs, As, Bs)

is stabilizable.

The decomposition of the system is carried out based on a variable parameter system in a multilayer structure using a directed graph (digraph) representation.

A one to one correspondence between the nodes of a cyclic digraph and a set S of variable parameter systems is considered and a digraph of variable parameters system is defined.

That is a hierarchical dependence of the variable parameter is established so that these variable parameters in a system S can only be a function of the states of those systems on edges running directly to S.

Another graph theory approach is presented in (Siljak, 1978) where an 'overlapping decomposition' scheme is shown whereby a dynamic system is decomposed into overlapping subsystems.

The dynamic system S is of the type

$$\dot{x} = f(x,u,t)$$
$$y = g(t,x) \quad x \varepsilon R^m, \quad y \varepsilon R^l$$

Associated with the system there is a digraph

$$D = (V,R) \qquad V = uUxUy$$
$$R: (u_j, x_i), (x_i, x_j), (x_j, y_i) \quad \begin{array}{l} U = u_1 \ldots u_m \\ X = x_1 \ldots x_n \\ Y = y_1 \ldots y_l \end{array}$$

S is decomposed into S interconnected subsystems S_i

$$\dot{x}_i^* = f_i(t, x_i, *u_i*) + h_i(t,x)$$
$$y_i^* = g_i(t, x_i^*) \quad i = 1 \ldots S$$

An overlapping condensation $D^* = (V^*, R^*)$ is a digraph where $V^* = u^*Ux^*Uy^*$

$$u^* = (u_1^*, \ldots u_p^*) \quad u_i^* \varepsilon R^{m_i}$$
$$x^* = (x_1^*, \ldots x_s^*) \quad x_i^* \varepsilon R^{n_i}$$
$$y^* = (y_1^*, \ldots y_q^*) \quad y_i^* \varepsilon R^{l_i}$$

$$n_i < n,$$
$$m = \sum_{i=1}^{p} m_i, \quad l = \sum_{i=1}^{q} l_i$$
$$x = \bigcup_{i=1}^{S} x_i, \quad u_i \cap u_j = \emptyset, \quad Y_i \cap y_j = \emptyset$$
$$\qquad\qquad\quad i \neq j \qquad\quad i \neq j$$

Information theory approaches

In (Conant, 1972) (Dufour and Gilles, 1978) a methodology for structural analysis and partition of dynamic complex systems is presented. The authors assumed it natural to partition the system into slightly coupled subsystems. In order to do this a system S is represented by a boolean graph. $G \{(x), (1)\}$ where (x) is a set of characteristic variables and (1) the set of edges that represent their relationship. The intensity of the relationship between variables is introduced in order to have a valuated graph.

This intensity measure is made by two procedures, one from the information theory and the other from the Fuzzy set theory.

The information theory approach:

Given the characteristic variables

$$Z = \{x_1, x_2 \ldots x_a, x_p\}^T$$

the observation of this system in a given time range leads to a set of data x_a^k being the kth observation of the x_a variable. No linear assumptions are made. A "transinformation coefficient" is defined which allows to measure the linkage intensity.

The informational entropy of a variable x_a is calculated from N observations.

$$H(x_a) = \log N - \frac{1}{N} \sum_{i=1}^{Ca} \eta_a(i) \log \eta_a(i)$$

where there are $Ia = \{1,\ldots i, Ca\} x_a$ variable classes and $\eta_a(i)$ is the number of events happening.

The conjoint entropy of two variables x_a and x_b is defined by

$$H(x_a, x_b) = \log N - \frac{1}{N} \sum_{i=j}^{Ca} \sum_{j=1}^{Cb} \eta(i,j) \log \eta(i,j)$$

A transinformation of x_a to x_b is introduced as

$$T(x_a : x_b) = H(x_a) + H(x_b) - H(x_a, x_b)$$

This is a measure of the relationship intensity between x_a and x_b.

For a complex system

$$T(x_1 : x_2 \ldots x_a : x_p) = \sum_{i=1}^{p} H(x_i) - H(x_1, x_2 \ldots x_p)$$

After that a static coupling matrix can be constructed. This matrix can be decomposed with the help of decomposition techniques of valuated graphs. This decomposition is then of static systems.

In the case of dynamic systems partition the same procedure is followed

$$\{x_a^1, x_a^2 \ldots x_a^{N-1}\}$$
$$\{x_b^2, x_b^3 \ldots x_b^N\}$$

$$T(x_a : x_b^i) = H(x_a) + H(x_b^i) - H(x_a, x_b^i)$$

So that the decomposition technique is the same for the static case.

The fuzzy theory approach:

Given a set of p characteristic normalised variables

$$Y = (y_i \ldots y_p)$$

the linkage coefficient between yi and yj can be expressed from the measurement of dissimilarity between yi and yj. To obtain a fuzzy image of yi two operations Q and Q^1 are carried out. Q for the space partition and Q^1 for the time partition. Having obtained a fuzzy image of yi → yj they are compared on the Hamming's generalised relative distance. A partition of the distance matrix is then carried out.

SYSTEM ANALYSIS

The second step in the design of a D.C.C.S. is the analysis of the model or set of models of the plant. The analysis considers the observability, controllability and stability studies of the model(s) of the plant.

Stability

The study of the stability of large scale systems already supposed a group of subsystems and their interconnection specification. If each one of the subsystems is stable and a measure of the coupling between subsystems is provided, then a first analysis would be how big or small should the magnitude of the coupling be in order to keep the system as a whole stable. Different studies that concern this aspect of stability are surveyed in (Sandell and Colleagues, 1978). Another class of stability that in large scale systems is being disregarded to a certain extent is structural stability.

Structural stability

In (Bernusson and colleagues, 1978) the study of the stability of interconnected systems subject to structural perturbations is presented, (i.e. interconnection pattern or subsystem structure variation) through a vector Lyapounov function analysis.

The components of the vector Lyapounov functions are in most cases, Lyapounov functions for the isolated systems.

The interconnection terms are assumed to satisfy inequalities where some characteristics of the Lyapounov functions of the isolated subsystems are involved. The large scale system is described by

$$\dot{x} = F(x,t)$$

and the subsystems $(S_i) i = 1...S$ are

$$(S_i) : \dot{x}_i = g_i(x_i,t) + h_i(x,t)$$

Stability conditions have been derived in the case when no predescribed pattern is given for the sequence of appearance of the structural perturbations. No assumptions on linearity are made. The perturbations are defined as a countable set of couples.

$$(G,H) = G = \{g_1(x_1,t)^T, ... g_s(x_s,t)^T\}^T$$
$$H = \{h_1(x,t)^T, ... h_s(x,t)^T\}^T$$

i.e. they act both on the isolated system s_i and the interaction h_i.

The problem is to check the stability of the equilibrium of S in the case when g_i, h_i functions take any form in time among a set P of allowable structural perturbations without any prescribed pattern. This is solved using the concept of connective stability.

Controllability

A linear continuous system is controllable under a decentralised information structure if there exists a decentralised control law which transfers any unknown initial state to the origin in a finite time interval.

In (Kobayashi, Hanafusa, Yoshikawa, 1978) Necessary and Sufficient conditions of the controllability under decentralised structure are given.

It is shown that controllability under a decentralised information structure is a necessary condition for pole assignability with local dynamic feedback control. Pole assignability for systems with two control stations is also studied. Signalling strategies are used to pass information to other stations especially if the control law is designed for the purpose of information transmission.

In (Aoki and Li, 1973) the authors study the controllability of an already decomposed system of the form

$$\dot{x}(t) = A x(t) + B_1 u_1(t) + B_2 u_2(t)$$

$$y_i(t) = H_i x(t) \quad i = 1,2.$$

They examine the controllability of all the system by a single agent, and study the interactions of the agent control actions.

The controllable subspace of agent $i < A/B_i >$ is not invariant with respect to various feedback modifications of A. The problem of possible reduction or enlargement of an agent's controllable subspace by another agent through properly choosing its inputs is studied in detail. The case when each agent knows entirely the state vector without error and when only a part of the state vector is known by the agent is also studied. They apply the results to obtain several necessary and/or sufficient conditions for stabilising decentralised systems by state feedback or by local output feedback signals. It is also shown that exchange of observation and control data co-operation between agents are important in achieving system stabilisation.

It is not unnecessary to say that stability and observability are also criteria for model decomposition and for evaluation of already decomposed models.

CONTROL SYSTEM DESIGN

We will now turn to the third step in the design of the D.C.C.S.: the synthesis of the control system.

The design problem in decentralised control is usually approached as an extension of present modern control solutions following one of the next two solutions schemes.

Let the system S be described as $S : (s_i, ... s_n)$ and their interrelationship.

Scheme 1 - Consider a set of "independent" subsystems s_i (i.e. neglect interconnections)

- Modify the solution obtained to consider interrelations.

In this solution scheme one considers first that only local information is available to each control function, and then considers some additional (non local) information.

Scheme 2 - consider the overall system S
- eliminate some information exchanges monitoring the performance degradation.

In this solution scheme one considers first that global information is available to all control functions and then considers some restrictions in information knowledge for the control functions.

More advanced approaches consider in addition to the classical centralised design criteria, the information pattern as a design criterion.

The control system can be what is known as decentralised or hierarchical. In both cases in industrial plants the states of the plant are not always measurable and noise free, so the first problem that arises is that of decentralised observation and decentralised estimation.

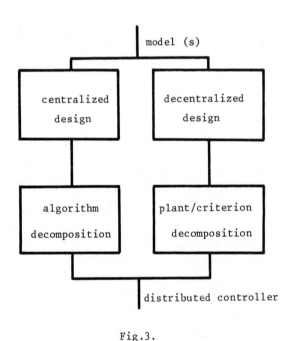

Fig.3.

Decentralised Observation

Simple and efficient state observation schemes for large scale systems described as interconnections of several low order subsystems are presented in (Sundaresham, 1976). The study is conducted within a decomposition decentralisation framework and consists of obtaining suitable modifications of a decentralised observation scheme comprising a set of local observers designed on the basis of the dynamics of the decoupled subsystems.

Each controller possesses partial and probably non identical information about the state of the system. The central idea of the analysis is the characterisation of the interconnections among the subsystems as perturbation terms acting in contradiction to the autonomy of the individual subsystems. The question: under what modifications do a set of local observers designed on the basis of the subsystem dynamics alone serve as an observer, with certain desired properties for the overall system? is answered.

Consider a large scale system S that may be described as an interconnection of s subsystems $s_1, s_2, \ldots s_s$.

$$\dot{x}_i(t) = A_i x_i(t) + B_i u_i(t) + h_i(t,x)$$

$$y_i(t) = C_i x_i(t) \qquad i = 1\ldots s$$

where

$$x_i(.) \varepsilon R^{n_i} \qquad x \varepsilon R^n$$

$$u_i(.) \varepsilon R^{m_i}$$

$$y_i(.) \varepsilon R^{p_i}$$

$h_i: R \times R^n \to R^{n_i}$ is a function that specifies the interconnection of s_i with the rest of the subsystems.

$A_i C_i$ are completely observable.

The overall system can also be described by

$$\dot{x}(t) = Ax(t) + Bu(t) + h(t,x)$$

$$y(t) = Cx(t)$$

The nonlinear function $h(t,x)$ describes the interconnection pattern within the overall system s.

if $h(t,x) = 0$

the observation scheme is then classical

$$\dot{\hat{x}}_i(t) = (A_i - K_i C_i)\hat{x}_i(t) + K_i y_i(t) + B_i u_i(t)$$

$$i = 1\ldots s$$

$$K_i = G_i C_i^T$$

where G_i is the positive definitive solution of the Ricatti equation

$$G_i(A_i^T + \eta I_i) + (A_i + \eta I_i)G_i - G_i$$
$$- G_i C_i^T C_i G_i + Q_i = 0$$

$$Q_i \geq 0$$

The problem is to seek modifications of this decentralised observer which result in the same degree of convergence η when $h_i(x,t) \neq 0$

When the interconnections are linear and time invariant

$$h_i(t,x) = \sum_{j=1}^{S} H_{ij} x_j(t) \qquad H_{ij} \in R^{n_i \times n_j}$$

The system may be described by the composite equation

$$\dot{x}(t) = (A + H)x + Bu \qquad u = Cx$$

if C and H satisfy

Rank (CH) = Rank (C) = p

The modified observation scheme is

$$\dot{\hat{x}}_i(t) = (A_i - K_i C_i)\hat{x}_i(t) + K_i y_i(t)$$
$$+ B_i u_i(t) + \sum_{j=1}^{S} \bar{K}_{ij} y_i(t) \qquad i=1\ldots s$$

where $\bar{K}_{ij}$ are elements of the matrix
$$\bar{K} = HC^T(CC^T)^{-1}$$

In comparison with the precedent observation scheme ($h_{ij}=0$) the only modification is the use of the outputs of the other subsystems as additional input signals to each observer, with the gains $\bar{K}_{ij}$ being determined by the matrices H and C.

Note that the computations of the gains K_i which determine the degree of convergence of the scheme is made at the subsystem level and more particularly the solution of a Ricatti equation for the overall system is avoided.

However note that the communication requirements between subsystem observers are critical.

Decentralised Estimation

Quite often in the control of a system it is necessary to generate estimates of the system state from noisy measurement data. Attention has been focused on state estimation techniques which are compatible with completely decentralised information structures and it has been shown that effective algorithms from this class can be developed.

In (Sanders and colleagues, 1976) the authors systematically explore some of the tradeoffs between the cost of implementing information transfer and the resulting performance improvement. The design and evaluation of large scale estimation algorithms which utilise information exchange over noisy communication channels is considered. An algorithm for the design of optimal information exchange is presented.

When it is not feasible to obtain complete measurements of the system interactions, two design techniques are presented in (Sanders, Linton and Tacker, 1978) for a specific filtering structure wherein information exchange between local filters is allowed for the purpose of obtaining estimates of the system interactions.

Decentralised Stabilizability and Stabilization

In (Aoki, 1972) the stabilizability of decentralised linear time invariant dynamical systems with co-ordination and/or communication among control agents is studied.

Consider a linear time invariant dynamic system with k local controllers (control agents, decision makers) described by

$$\dot{x} = Ax + \sum_{i=1}^{k} B_i u_i \qquad x \in R^n \quad u_i \in R^{r_i} \quad i = 1..k$$
$$\dim B_i = (n \times r_i) \qquad \sum_{i=1}^{k} r_i = r$$

Each local controller is assumed to have its own observation scheme

$$y_i = H_i x(t)$$

where y_i is the m_i dimensional observation vector

$$\sum_{i=1}^{k} m_i = m$$

Two types of controllers may be considered

- Static

$$u(t) = N(t)y(t) + v(t)$$

$$y^T(t) = (y_1^T(t) \ldots y_k^T(t))$$

$$u^T(t) = (u_1^T(t) \ldots u_k^T(t))$$

$$v(t) = (v_1^T(t) \ldots v_k^T(t))$$

- Dynamic

$$\dot{s} = Ls + My$$

$$s^T = (S_1^T \ldots S_k^T)$$

$$u(t) = Ks(t) + v(t)$$

$$\dim(S) = 1 \quad \dim(s_i) = 1_i$$

Imposing different structural constraints on matrices K,L,M,.N different information patterns imposed on the decentralised system can be modelled. In addition to the variety of information patterns and the interactions of local controllers for the decentralised control problems, the observation schemes could be used to classify the decentralised systems.

- independent observation where for S_i $i=1 \ldots k$ are all independent
- dependent observation where $\bigcap_{i=1}^{k} S_i \neq \{0\}$

The close loop system with dynamic controller is

$$\begin{bmatrix} \dot{x} \\ \dot{s} \end{bmatrix} = \begin{bmatrix} A & BK \\ MH & L \end{bmatrix} \begin{bmatrix} x \\ s \end{bmatrix} + \begin{bmatrix} B \\ 0 \end{bmatrix} v$$

and with static controller is

$$\dot{x} = (A + BC)x + Bv$$

In centralised control when no structural constraints are imposed under C, the controllability of the pair (A,B) is equivalent to the pole assignability so that stabilizability is equivalent to controllability of unstable modes of A.

In the case of decentralised control the C matrix is structurally constrained so that stabilizability is no longer equivalent to controllability of unstable modes.

In the case of dynamic controllers the characteristic equation of the closed loop system is

$$0 = \begin{vmatrix} A-\lambda I & BK \\ MH & L-\lambda I \end{vmatrix}$$

$$(L-\lambda I)^{-1} = (L_1-\lambda I)^{-1} + (L_2-\lambda I)^{-1} + \ldots + (L_k-\lambda I)^{-1}$$

The problem of stabilizability is solved for two special cases:

- when A is block lower triangular there always exists L and v^T where v_i is generated by the controller i as a proportional feedback for any choice of M and K that stabilise the system with no communication among agents.

- when B, H, L, K and M are diagonal matrices then the overall system is stabilisable.

In the case of static controllers if the pair (A,B) is completely controllable with

$$R^n = \{b, \ldots A^{h_1-1}b_1, \ldots b_t, \ldots A^{h_t-1}b_t\}$$

where $t \leq n$

Then the system is stabilizable by the output feedback with the observation scheme

$$\text{where } H = \begin{bmatrix} H_{11} & 0 & \ldots & 0 \\ 0 & & & \\ \vdots & & & \\ 0 & 0 & & H_{tt} \\ & H' & & \end{bmatrix}$$

$$u_i = d_i^T y_i \quad i=1 \ldots t$$

$$u_j = 0 \quad j = t+1, \ldots r \text{ if } t < n$$

Until now, no explicit communication among control agents has been introduced. If certain information exchange takes place between agents, the system is stabilisable from the output feedback under certain conditions. The case, where the R^n state space is decomposed as $R1 \oplus R2$ and agent 1 controls $R1$ and agent 2 controls $\hat{R}2$ is studied.

The exchange of information can be done between agents or through a coordinator level. A cost of communication must be introduced to be able to decide on an optimal division of Ro.

The subspace Ro which is the controllable subspace of both agents, requires either communication and co-ordination of control actions by both agents or assignment to one of the agents or by a central agent. The subspace Ro varies with partial feedback.

Finally cost for communication, more precisely of transmitting finite dimensional vector quantities among agents and cost for computation must also be incorporated.

The problem of stabilising a system by using local feedback controllers and some limited information exchange among local stations is considered in (Wang and Davison, 1978). The problem of achieving a given degree of stability with minimum transmission for linear time invariant system is solved.

Decentralised Pole Assignability

In (Corfmat and Morse, 1976) concepts from graph theory and from geometric theory of linear systems are used to derive explicit conditions for determining when the closed loop spectrum of a multichannel linear system can be freely assigned or stabilised with decentralised control. The authors study the effects of decentralised feedback on the closed loop properties of the k channel, jointly controllable, jointly observable linear systems. Channel interactions within such systems are usefully characterised in qualitative terms by means of suitable defined directed graphs. Complete systems prove to be precisely those systems which can be made both controllable and observable through a single channel by application of nondynamic decentralised feedback to all channels.

It is shown that if the graph of a k channel system is strongly connected, then the closed loop spectrum of the system can be freely assigned when the system is complete.

It the graph of the system is not strongly connected then the spectrum assignment is possible if and only if (1) the dimension of the system equals the sum of the dimensions of a set of strongly connected subsystems of the system and (2) each subsystem is complete.

If, for a particular system, free spectrum assignment is not possible then the spectrum of the closed loop system which results from the application of any specific linear decentralised control, contains a uniquely determined subset which cannot be shifted no matter what control is applied. By characterising this subset directly in terms of the system's parameters, a complete solution is obtained to the problem of stabilising a k channel linear system with decentralised control.

Decentralised Optimal Control

In decentralised optimal control an optimal control strategy is obtained by minimization of a criterion. Generally the system is considered to be linear and of the form

$$x(i+1) = Ax(i) + \sum_{i=1}^{k} B_i u_i$$

$$y(i) = Cx(i) + \sum_{i=1}^{k} D_i u_i$$

Note that the system is already decomposed so that the control actions are localised.

The criterion is of the form

$$J = \sum_{i=1}^{\infty} [x^T(i)Qx(i) + u(i)^T Ru(i)]$$

It is important to note that the decentralization we are referring to here is of the implementation of the control law.

We will see in the next section of Hierarchical Control that a further decomposition of the criterion is also called decentralised control but in this case is the computation (off-line) of the control law that is decentralised. We will see the implications of this later.

We will generally call on line decentralization the schemes which lead to actual decentralised control strategies, regardless of the object that has been decomposed; the system; the criterion or the algorithm. An offline decentralization is when the computation of a control law (centralised or decentralised implemented) is carried out in a decentralised fashion.

To rule out any confusion I will remind you that in Large Scale System Optimization because of the size of a problem (MP) (master problem) one is obliged to decompose it in order to have a manageable size of problems (auxiliary problems (AP)). The solutions of the auxiliary problems constitute the solution of the master problem. Nevertheless the solution in an optimal control problem is a centralised-control law. Decentralization is concerned with reduction of the online communication and computation requirements implicitly in a mathematically defined control law. Decomposition is concerned with reduction of the offline computations required to obtain the given control law.
Decentralization can be achieved without use of a decomposition method (in the optimization sense) at the expense of increased off-line computation. On the other hand a decomposition method can be used to compute a completely centralised solution.

Combination of the two methods holds out the possibility of reduction in both on-line communication and off-line computation.

In (Aoki, 1971) optimization problems for large dynamic systems with several control agents are considered. This class of problem differs from the usual large optimization problem in that no single control agent (data processing device) possesses a complete description of the total system and transfer of such dispersed information to a central location is either unfeasible or impossible. Two or more control agents must co-ordinate their control actions, either through mutual communication or via higher level control agents in order to achieve the optimal control of such decentralised systems.

Data processing locations must co-operate in solving a given optimisation problem by sharing their knowledge in system dynamics and intermediate computational results. This need for co-operation in turn gives rise to associated optimisation problems involving obvious tradeoffs among such design factors as the amount of information to be communicated, quality of control, data processing capacity at local or higher level etc.

In (Linton and colleagues, 1976) a computational approach to the problem of determining the optimal constant feedback gain matrices for a particular decentralised controller is presented.

Decentralised control has been applied to Packet Switched Satellite Communications (Schoute, 1978).

Hierarchical Control

We have pointed out before that most of the literature in decentralised hierarchical control refers to centralised implementations of the actual control law although their calculation off-line is done in a decentralised fashion to reduce complexity. We present some examples of these approaches.

Another less common approach is the hierarchical implementation of control systems that may or may not use hierarchical decentralised control schemes; these approaches are called multi-layered structure (Sandell and colleagues 1978) where the determination of control is split into algorithms which operate at different time scales. In general these approaches do not go further than the description level.

Multilayer control

In (Williams 1978) the structure of a general hierarchic (multilayer) system is outlined. A very detailed specification of the functions to be performed at each level and their relationship is presented. The feasibility of a hierarchic control structure from the available technology point of view is also studied. Tasks and Duties for each one of the levels are described. The assignment of personnel to some tasks is also presented.

One of the rare multilayer approaches that goes as far as the implementation is presented in (Binder and colleagues 1977). The organisation of a hierarchical control system for a complex process is proposed and illustrated by its application on a pilot distillation process. A co-operative organization of the control system is partitioned between several levels. The system is designed to satisfy the possibility of actual distribution of the computer support and the progressive realisation.

Hierarchical control

In Singh and Hassan 1977) and (Hassan and Singh 1976) a two level method is presented to determine the optimal open loop control that minimises the separable cost function.

$$J = \sum_{i=1}^{N} \int_0^T \tfrac{1}{2}(||x_i||^2 q_i + ||u_i||^2 r_i)dt \quad \text{for}$$

the non linear system given by:

$$\dot{x} = f(x,u,t)$$

This method uses an expansion around the equilibrium point of the system to fix the second and higher order terms. These terms are compensated for iteratively at a second level by providing a prediction for the states and controls which form a part of the higher order terms.

A co-ordination – decomposition approach is used. The control law is centralised and open loop. However, it is obvious that a centralised control of this type can be made on line distributed control considering a finite horizon and internal model controller structure.

In (Singh and Hassan 1978) the previous hierarchical optimisation algorithm for non linear interconnected dynamical systems with separable cost function is extended to the case of non linear and non separable cost function. This ensures that any decomposition (of the criterion) could be used and makes the algorithm suitable for the optimisation of general non-linear problems.

In this other algorithm the given desired state and

control trajectories are used to make the cost function a separable quadratic function by adding certain terms, and the dynamics linear. These desired trajectories are improved at the second level in order to force them towards the optimal states and controls.

In (Singh and Titli, 1978) after a review of the open loop approaches that have been presented above, a closed loop hierarchical control for linear quadratic problems is presented.

Decision and Game Theoretic Approaches

A different approach to the decentralised optimal control through decision theory has also been developed. The decentralised optimal control problem is usually stated in team decision theory as:

(Ho and Chu, 1972), given a team composed by $i \in I = (1,2...N)$ members. Each member receives certain information z_i and controls the decision variable u_i. The payoff function of all the members is:

$$J = J(\gamma_1, \gamma_2, \ldots \gamma_N)$$

where γ_i is the control law or the decision rule.

$$u_i = \gamma_i(Z_i)$$

$\gamma_i \in \Gamma$ Γ is the class of admissible controls.

The problem is then to find γ_i s.t. $\forall i$

$$J(\gamma_1, \gamma_2, \ldots \gamma_N) = \min.$$

Now we will define two classes of teams problems: static, dynamic.

We said that z_i is the information available to member i or decision maker DM_i in the team. This information z_i can have different structures.

Let $\xi \in R^n$ be a random vector of Gaussian distribution $N(0,X)$ with $X > 0$.

The information z_i is assumed to be a known linear function in R^{q_i} if ξ and some of the control actions other members have taken.

$$z_i = H_i \xi + \sum_j D_{ij} U_j \quad \forall i$$

H_i, D_{ij} are matrices known to all the members.

Real systems are causal so that what happens in the future cannot affect what is observed now. i.e. if the control action j affects the information i then u_i cannot affect the information j. In a causal system there is a partially ordered relation existing among the DM's.

$$D_{ij} \neq 0 \Rightarrow D_{ji} = 0$$

In general a static team problem (Ho, Kastner and Wong, 1978) is defined when the information z_i available to the DM's depends only on ξ

$$z = \eta \xi$$

η is known as the information structure.

When different DM's act at different times, the information z_i received later by DM_i may be dependent upon the action u_j of DM_j who acted <u>earlier</u>. In this case the decision problem is known as the dynamic team problem. "Dynamic" indicates the presence of order of actions of the DM's.

$$Z = \eta(\xi,u)$$

η must satisfy the causality conditions.

The information structure can be graphically represented by a precedence diagram with 'memory communication lines', where each Decision Maker is represented by a node. It is very important to note (Ho and Chu, 1974) that if one person acts more than once at different times, he is considered to be made up of several DM's each acting once and only once. So that the meaning of a decision maker is more a decision made regardless of who has made it.

It is important to note that the system description state-variables and dynamics (model) play a secondary role in this approach. Their effects are incorporated as part of the specification of the information structure.

In (Ho and Chu, 1973) a sufficient condition for the equivalence of dynamic and static information structures is discussed.

From the view point of decision and control theory, the purpose of acquiring information is to produce a better payoff.

Hence most simple definitions of the value of information are payoff dependent. Consider a set of decision makers with a common payoff function $J(u,\xi)$. The value of information is defined:

Value of information = the best you can do with the given information - the best you can do without that information.

The expected value of information (EVI) is simply the above averaged over all possible "state of the world", i.e. the random variables ξ

$$EVI = \max_{\gamma \in \Gamma} E(J(\gamma(z),\xi)) - \max_{u \in U} E(J(u,\xi))$$

The principal use of this definition is the comparison of different information structures for a problem. The η information structure is said to be more informative than η' if

$$EVI_\eta - EVI_{\eta'} > 0$$

An attempt of unifying results from three separate subject matters, namely, team decision theory, classical Shanon information theory and market signaling is present in (Ho, Kastner and Wong, 1978).

Centralised Control, Algorithm Decomposition

Another approach for designing distributed control systems is to decompose centralised control algorithms, so that they can be executed in different computational stations. This approach is used when time critical or large scale systems pose a problem of computational speed or burden. Furthermore some control algorithms that were impossible to use because of the numerical amount of computation could now be implemented using distribution of the computational load. Real time numerical analysis is now possible, because of parallel or concurrent numerical analysis. Increasing literature on parallel numerical analysis is now available.

Functional decomposition of centralised control algorithms is the most natural since control blocks have functional meaning. However different levels of decomposition could arise; at the control function level, the mathematical function level, mathematical operation, instruction, etc.

In (Nader 1979) a Time Valued Petri Net decomposition approach is presented.

This approach allows one to compare different decompositions and to estimate the minimum execution time to compute a control action.

The tradeoff of management requirements and parallelism introduced by several decompositions is also studied.

Centralised control algorithms decomposition can be considered in a more general framework of software decomposition. Software decomposition is generally approached through graph or net theory techniques. One associates to a program (in this case the control algorithm program) a graph or a net (e.g. data flow graphs or control flow graphs). The techniques used can be identified as segments, DD paths, intervals, classes etc.

II

THE COMPUTER SYSTEM

DISTRIBUTED PROCESSING

The concept of distributed processing has been in computing science for more than a decade. However there is not yet a well established definition of distributed processing. Many attempts have been made to clarify its meaning. (Desautels, 1978) (Infotech, 1976), (Infotech 1977).

(Ravi, 1976) establishes a categorization of distributed systems. This categorization is based on the data known by each computational agent. If a function is distributed and each agent has access to all sets of data manipulated by all agents for that function, the agent is said to have global knowledge. If this is not so and each agent only has access to its own set of data, then it is said to have local knowledge; any degree of access between these two is called partial knowledge.

(Casaglia, 1976) defines distribution following two different approaches: physical distribution and logical distribution. Physical distribution corresponds to 'function distribution' over different processors or computers. Logical distribution corresponds to 'control distribution' over different local or remote processors. Following this approach a multi-processor is a distributed system.

(Enslow, 1978) establishes a framework for discussing a variety of objectives. This definition considers three basic dimensions: hardware organization, control organization and data base organization. A region of this space is defined as the allowable region for distributed processing.

Our definition that intends to be less technologically oriented concentrates on the logical aspects of processing rather than on the implementation aspects.

We define a process as a quintuplet

$$P = \{I, O, CX_o, CX, A\}$$

where

I are the data inputs of P
O are the data outputs of P
CX_o is the initial state (context) of P
CX is the state (context) of P
A is the algorithm definition $I \xrightarrow{A} O$

This process definition implicitly assumed a

sequential processing of A. To generalise this definition we explicitly specify the control strategy assumed.

$$P = \{I, O, CXo, CX, A\} \cup \{Control\}$$

where

$$\{Control\} = \{Process\ Communication,\ Process\ Synchronization\}$$

i.e. the specification of the process interrelations with other processes in terms of execution order and sharing of information.

We say we have a distributed system S when:

$$S = \{\bigcup_{i=1}^{n} P_i\} \cup \{Inter\ P_{ij}\ Synchronization,\ i \neq j\ Inter\ P_{ij}\ Communications\}$$

Based on this definition and on that of D.C.C.S. it is easy to realise that already developed areas in multiprocessing are relevant to D.C.C.S.

CONCURRENT PROGRAMMING

Concurrent programming in D.C.C.S. has the combined difficulty of distributed and real time systems. Although there is no well established methodology even in centralised computer control system design it seems that the preparadigmatic stage is at its end with the convergence of certain design concepts. In computer architecture and operating systems design, there are concepts that are being widely accepted.

The Layered Structure

At the overall design strategy level the hierarchical (layered structure) conception is generally accepted, although not everyone explicitly specifies which is the order relation in the hierarchy and sometimes this relation is not constant through the designed layered structure. This strategy together with the separation of specification, design and implementation are key concepts in the overall design of distributed computer systems.

The idea of structuring a program in modules that make use of simpler modules to perform its function goes back to the beginning of programming languages. However the formal methodology for hierarchical systems was first established by (Dijkstra, 1968) and (Brinch Hansen 1970). The order relation in the hierarchy for Hansen's approach is that of property or privilege transmission. There have been many applications of these approaches to the design of operating systems.

The Module Concept

The next essential design concept that has been widely accepted is that of the 'module'. The hierarchical design strategy has forced the designers to establish suitable definitions of 'modules' and their interrelationship. The need for a clear specification of modules and their interconnection has been satisfied with the association of protection mechanisms to modules. Reliability considerations have also introduced the use of assertions in these modules.

The 'module' concept is as old as that of the hierarchical systems. Hierarchical systems are always formed by well defined entities called modules. The essence of the module concept is its easy semantic interpretation. Logic independence with other modules provides a sound basis for insurance of protection and correctness. The concept of information hiding (do not let others know how your module is designed in order to avoid dependency of theirs on yours) is presented in (Parnas, 1972a). Parnas (Parnas 1972b) discusses the criteria to be used in dividing a system into modules. The modularization includes the design decisions which must be made before the work on independent modules can start. Modules can be considered as a responsibility assignment.

Interprocess Communication and Synchronization

Inherent problems to the 'module' concept in real time, multiprocessing or distributed systems are the interprocess communication and synchronization. Here again we observe a convergence of the basic concepts and a clear separation between specification and implementation. A review of the most common synchronization and communication mechanisms is presented in (Lucas, 1975), (Lister, 1977).

The most widely accepted mechanisms are:

- flags
- semaphores
- message passing
- D operations and excluding regions
- critical regions and conditional critical regions
- monitors
- path expressions

Protection and Correctness

The first level of protection is always given by the syntax of the software language. When further protection mechanisms are required the most used are:

- capabilities
- rings

- ports
- segments
- domains

The introduction of assertions in the module guarantees at the design level that the operation of the module corresponds to its specification.

Because of the relative interpretation of a system as a set of processes at different levels of abstraction, the above general design concepts apply to any of these levels of abstraction from the system to the lowest application level.

Some Approaches

The different combination of the above ideas applied in distributed systems produced many different approaches. We mention some of them here.

Brinch Hansen (1978) introduces the language concept of 'distributed processes' without common variables. These processes communicate and synchronize by means of procedure calls and guarded regions (Dijkstra, 1975). This concept is proposed for real time applications controlled by microcomputer networks with distributed storage. The author claims that procedures, coroutines, classes, monitors, processes, semaphores, buffers, path expressions and I/O are special cases of 'distributed processes'.

Hoare (1978) suggests that I/O are basic primitives of programming and that parallel composition of communicating sequential processes is a fundamental program structuring method. He also uses Dijkstra guarded commands.

Kramer and Cunningham (1978) present an approach to the design of distributed processing. Systems are described as a set of logically concurrent asynchronously interacting modules. Modules are finite state machines which respond to input changes to achieve their contribution to the function of the system. The design includes consistency checks at the system and module levels. The approach is based on the informal use of verifying assertions in a programming notation.

DISTRIBUTED OPERATING SYSTEMS

The design of distributed operating systems uses extensively all the concepts mentioned above. When in a distributed system for process control, multiprocessing is not allowed in each processor; the design of a distributed real time operating system is straight forward because it is formally equivalent to a multiprocessing real time centralized operating system where virtual resources and concurrency become real. When multiprocessing is allowed in each processor of the distributed system, there is a double level of multiprocessing: the distributed system and the single processor. In this case the design of a distributed real time operating system is more difficult.

Some Examples

The earliest serious approach to the design of a multiprocessor operating system seems to be Hydra (Wulf and colleagues, 1974). Hydra is the kernel of a multi processor. The notion of resource (physical or virtual) has since then been called an 'object'. Mechanisms are presented for dealing with objects: creation of new types, specification of new operations, sharing and protection of any reference to a given object against improper application of any of the operations defined with respect to that object. Extensions of the system are easily accomplished with these mechanisms. The hierarchical structure is not followed in this system. The system was implemented for a 16 processor architecture connected through a cross bar switch.

Roscoe (Solomon and Finkel, 1978) is perhaps the most recent multi-microcomputer operating system. This distributed system runs on loosely coupled identical processors without shared memory (5 LSI-11). The fundamental unit of execution is a process. Each processor runs one or more processes. Processes are constrained to run at one processor at a time although migration is possible in order to balance the computational load. The communication among processes takes the form of messages. The fundamental protection mechanism is a 'link'. Links are permissions and pathways for messages. They are given selectively only to processes that are to have the type of access that the link provides. A flow control algorithm is used to allocate the scarce system resources of message buffers and inter-machine links.

Farber and Larson (1972a,1972b) describe a software system (DCS) which allows the control of a network to be distributed among the processors of the network. The system is viewed as a single geographically distributed facility (i.e. one level of multiprocessing), rather than a set of individual processors interconnected. Responsibility for resource allocation and scheduling is distributed among the separate processors. The communications network is an unidirectional data ring. The system has a layered structure. The communication is through broadcasted messages. Messages are addressed to processes and not to processors.

Cm* (Jones and colleagues 1977) is a distributed multi processor composed of computer modules each consisting of a DEC LSI, a standard LSI 11 bus, memory and devices. The system provides sharing of code and data via a capability addressed virtual memory. A module is a 'unit of abstraction'. The kernel software supports the notion of a module by providing user facilities to create modules and to invoke functions of a module in a protected way. Module boundaries are used for protection purposes at run time. Modules may be partitioned into strictly ordered sets of levels. Multiple levels of one module share data structures and even code. The creation and management of groups of processes ('task forces') is provided by the system. The members of a 'task force' synchronise and communicate with one another through mailbox mechanisms.

The distributed computer network (Mills 1976) (DCN) is a resource sharing computer network which includes a number of DEC PDP-11's. DCN supports a number of processes in a multiprogrammed virtual environment. Processes can communicate with each other and interface with this environment in a manner which is independent of their residence within a particular computer. Resources such as processors, devices and storage media can be remotely accessed and shared so as to provide increased reliability, flexibility and system utilization. DCN processes are named and given a unique identity within the network (as in DCS). Most DCN operations are performed in the context of a process. The kernel includes mechanisms for I/O device interface, storage allocation, process scheduling, interprocess communication, etc.

A domain oriented distributed computer system is presented in (Casey and Shelness, 1977). The system components are domains, segments and virtual processors. The information necessary to represent and manage a virtual processor is a special segment called 'virtual processor segment'. A computation is effected by the progress of a virtual processor through a number of domains. As a computation proceeds different resources and services are required. The virtual processor moves to other domains by invoking an interdomain jump so as to be able to access the appropriate code and data. A change of domain is associated with a change of segment from which instructions are being executed. The inter domain jump is distributed. Distribution of functions is determined dynamically by the domain management mechanism. Fine grain load balancing is possible because load balancing decision is taken every inter domain jump.

Scheduling

Real time distributed scheduling can also occur at two levels: at the distributed system level and at the processor level. If multiprogramming is not allowed in each processor of the distributed system, the problem of real time scheduling is reduced to the problem of static decomposition, static assignment and static load balancing of processes. If multiprogramming is allowed in each processor of the distributed system, then dynamic decomposition, dynamic assignment and dynamic load balancing are possible although not necessary. Non pre-emptive schedules allow dynamic load balancing and dynamic assignment. If multiprogramming is allowed in each processor without dynamic decomposition, assignment and load balancing, then the problem of scheduling in the distributed system is similar to that in the centralised system.

In D.C.C.S. the absolute flexibility or dynamic restructuring is not likely to happen. In D.C.C.S. the computation locations must contain the application programs relevant to that location either geographically or functionally, or even more physically constrained because of hardwired connection of sensors and actuators to that location. Some flexibility is required in order to provide graceful degradation (Schoeffler, 1978) (Spang, 1976).

A unified formulation for lower bounds on the minimum number of processors and on time for multi processors optimal schedules is presented in (Fernandez and Bussel, 1973). A methodology based on a PERT approach is presented to estimate the minimum number of processors that allow one to execute a set of computations in a time not greater than the length of the critical path of the graph associated to these computations. Similarly they present an approach to determine the minimum execution time when the number of processors is fixed. The scheduling strategy is to execute those tasks, which predecessors have completed execution, in the first available processor. A similar approach that presents additional analysis features is presented in (Nader, 1979b).

A static assignment procedure in order to maximise the performance of distributed programs is presented in (Stone, 1978) (Stone, 1977). In a two processor distributed computer network a flow algorithm is used to determine the optimal assignment. The load of one processor is held fixed and the other varied. For every program module M there exists a critical load factor fM such that when the load on the processor with variable load is below fM, M is assigned to that processor by an optimal assignment, and is otherwise assigned to the other processor. This critical load factor opens the possibility of doing optimal dynamic assignments in real time. This is an elementary feedback control scheme. The model assumes that the cost to be minimized is proportional to total processor run time, plus the cost of communication from computer to computer when modules that are not coresident must transmit information across a communication link. To perform dynamic assignments of modules in an optimal fashion one needs only to determine the load factor of the computer at a given time and compare this to a list of precomputed critical load factors. An attempt to extend and generalise these ideas for the case of three or more processors is presented. This approach suggests that there may exist some relatively crude methods for measuring and estimating 'key variables' in real time to obtain approximations to the critical load factors and the actual load factor for on line dynamic assignments.

There have been relatively few studies which have attempted systematic analysis of schedulers which consider integrating feedback information into the operating system schedulers. (Brice and Brown, 1978) present some integrated feedback driven scheduling systems for multi programmed multi processor computer systems. Two basically different procedures for coupling CPU, I/O and memory schedulers are analysed for a spectrum of model system parameters. Dynamic feedback scheduling gives improved performance over static scheduling, particularly under conditions of heavy I/O processing demands. The rate at which an executing program would move data to and from executable memory if allowed to run unimpeded is possibly its most fundamental intrinsic characteristic. This characteristic is called the data flow rate, or I/O rate. The idea of using I/O rates as a basis for CPU queue priority is an old one. Processor demands are commonly defined by buffer sizes, buffer service algorithms and the access rates of rotating memory. Data flow rate of an executing process is used as a measure of its processor demand, both for I/O and CPU services instead of using buffer sizes. The data flow rate is then used to schedule I/O system resources, buffer sizes, CPU resources and service algorithms. The data flow rate is used as the control variable in the integrated feedback schedulers. It is the coupling of the fundamental data flow rate back to the CPU and I/O schedulers which allows implementation of dynamic feedback schedulers. Only a single easily gathered data item is required to define the feedback loop. This makes applications of this type of dynamic scheduling a low overhead process. Note that most major multiprogramming operating systems already gather data on I/O activity for accountancy purposes and for performance analysis purposes.

Two feedback algorithms that were implemented in the distribution-driven models are presented. The first attempted to give preference to processes which have files assigned to under utilized devices. This is called 'device-oriented feedback algorithm'. The second algorithm selects processes with a high data flow rate and allocates additional I/O processing capability to these processes. This is called 'process oriented feedback algorithm'.

The resource management techniques and algorithms suggest that the use of an intelligent and flexible processor, for the servicing of I/O requests is significant. This has deep implications for the architecture of future generations of computers.

The control theoretic approach for computer systems performance improvement is emphasised in (Jain, 1978). This paper supports control theory as a tool for dynamic optimization of computer system performance. For a control theorist an operating system is a set of controllers which exercise control over the allocation of some system resource. The objective of each controller is to optimise system performance while operating within the constraints of resource availability. Most analytical models of computer systems used at present are queueing theoretic based. In this case an operating system is interpreted as a server. Queueing theory is limited in its scope due to its steady state nature. Queueing theory is a good design and analysis tool for computer performance. However it provides little run time guidance. Control theoretic approaches are best suited for dynamic (on-line guidance) performance optimization. They are in addition design and analysis tool (Controllability, stability). A general control theoretic methodology for resource management is presented:

- modelling (stochastic models)
- parameter identification
- definition of the control law (resource management policy)

The CPU management policy is presented as an application example.

Computer Control Systems have now been developed for a long time. Time has come to develop Controlled Computer Systems.

APPLICATION SOFTWARE

For the design of application software in D.C.C.S.

- software decomposition and assignment,
- software requirements specification and software reliability and correctness

are essential issues.

Software Decomposition and Assignment

This design problem has been approached using language, graph and analytical tools. Decomposition using language concepts has already been presented.

Graph approaches

Graph approaches are mostly used at the application software level. They are less rich than the language ones, although used as an analytical tool, they are more efficient with the possibility of quantification of some aspects of the design. Traditionally graph approaches have been the most used for flow analysis. Baer (1973) reviews different models for parallel computation: graph models, Petri Nets, parallel flowcharts and flowgraph schemata. The predominant approaches used for partitioning program graphs are: (Paige, 1977)

- segments
- DD paths
- intervals
- classes
- level i paths.

Analytical approaches

Firestone (1977) present a model to evaluate the optimal number of processing units for multiprocessing a program with minimum cost. The optimal number of processors is evaluated through minimization of a cost criterion. For given computer and elapsed time cost factors there is a maximum number of processors that can be used effectively. Memory interference is taken into account in the criterion. A lower bound of the elapsed to computer times ratio below which parallel processing is not more effective than sequential computation is established. This lower bound depends only on the behaviour of the degradation factor. It is shown that the degradation factor is linear in the number of processing units employed, inversely linear in the number of independant storage banks and quadratic in the relative accessing rate.

Requirements Specification

Specifications provide the fundamental link to make the transition between the concept and definition phases of the system development process. Unambigious specifications are required to ensure successful results and at the same time minimize cost overruns during the development process. Many problems currently being addressed by software engineers have their origins in the inconsistent and incomplete nature of system specifications (Belford and Bond, 1976). This problem becomes more critical for distributed systems where the interrelation specification of the system has to be complete and consistent.

Ross and Schoman (1977) present an approach (Structured Analysis) to requirements definition. A systematic methodology that consists of context analysis, functional specification and design constraints is presented. Languages such as SA and PSL have been developed for requirements documentation and preparation of functional specifications (Ross, 1977) (Teichroew and Hershey, 1977).

A decomposition methodology that assists designers in identifying data processing requirements that are essential to meet the system requirements is presented in (Salter, 1976). System requirements are transformed into functional structure and operating rules. A methodology for the requirements specification of real time systems is presented in (Gaulding and Lawson, 1976) (Alford, 1977).

Reliability and Correctness

A reliable behaviour of a program can be expected only if it meets its specification. Assertions, Petri Nets and their combination are verification tools for correctness of parallel programs.

A generalized inductive assertion method for proof of correctness for multiprocess programs is presented in (Lamport, 1977). Flon and Suzuki (1978) present a consistent and complete axiomatization of total correctness (including safety from deadlock and starvation) suitable for automatic verification.

The Petri Net approach is very useful for the correctness proof of the control structure of concurrent programs. Szlanko (1977) presents a deadlock detection method using Petri Nets. Kellel (1976) uses a Net based tool ('conceptual model') for parallel program verification.

COMMUNICATION NETWORKS

When a D.C.C.S. is implemented in a computer network, its performance is directly affected by the performance of the communication network. Many solutions have been proposed to improve the performance of the communication network using: heuristic algorithms; analytical techniques such as optimization and queueing networks; and control theory. The essential objective been to minimize the average communication delay. Minimization of this delay is achieved through optimal flow control and routing, and minimization of contention. Other characteristics of the network such as error control and protocols also determine its performance. The influence of error control and protocols is not reviewed here. (Kimbleton and Schneider, 1975). The heuristic and analytical approaches are essentially useful at the design stage; they are static (Chou, 1978),(Trakhtengerts and Berezovsky, 1976).

A dynamic approach for performance improvement can only be achieved using control theory. The communication network is well suited for the implementation of its decentralized control. Allocation, routing and flow decisions must be mostly based on traffic characteristics known at each node.

Some attempts have been made to use control theory for the optimal control of dynamic routing. Feedback solutions to the linear optimal control problem with state and control variable inequality constraints for the message routing problem is studied in (Moos and Segal, 1976). A bang-bang control is found to be optimal. Meditch (1978) presents a multivariable pole assignment approach to the control of store and foreward communications network. Feedback routing policies which regulate queue lengths throughout the network in presence of heavy traffic are obtained. The control scheme is centralized and requires full state feedback. A controllability analysis of the data network is also presented.

Schoute (1978) presents a decentralized optimal control policy for access control to a packet switched network. He shows that the optimal control law for new packets is to have green light for all stations all the time (all together policy) or to have green light for each station in turn (round robin). For collided packets it was found that it is optimal to have predetermined assignments of groups of stations to time slots for retransmissions and a dynamic programming algorithm for determining the optimal grouping is given.

CONCLUSION

We have reviewed some of the design problems of D.C.C.S. We emphasized the approaches that use information sciences tools in the design of the distributed control and the approaches that use control theory in the design of distributed computer systems.

The basis of a general approach lies in the relationship between automata, discrete and continuous systems. The process and automaton concepts are strongly related. Automata are a class of discrete systems. The possibility of representing a continuous system as an automaton and vice versa opens the possibility of treating some problems of one field with the methodology of the other.

We believe the conceptual unity displayed here among these areas is new and hopefully will motivate new ideas and will lead to future cooperative effort among researchers in these fields.

REFERENCES

Alford M.W. (1977). A requirements engineering methodology for real time processing requirements.
IEEE Trans. Software Eng. Vol.SE-3 No.1, January 1977.

Aoki M. (1971). Decentralized control of large dynamic systems and a new class of associated optimization problems.
IFIP Congress on Information Processing. Ljublyana, Yugoslavia, August 1971.

Aoki M. (1972). On feedback stabilization of decentralized dynamic systems.
Automatica, Vol.8, pp.163-173, 1972.

Aoki M. and M.T. Li (1973). Controllability and stabilizability of decentralized dynamic systems.
JACC 1973, pp.278-286.

Baer J.L. (1973). A survey of some theoretical aspects of multiprocessing.
Computing Surveys Vol.5, No.1. March 1973.

Belford P.C. and Bond A.F. Specifications, a key to effective software development.
2nd International Conference on Software Engineering. San Francisco, California, 1976.

Bernusson J., Buryat G. and Bitsoris G. (1978). On the stability of large scale interconnected system under structural perturbations.
IFAC 1978 World Conference.

Binder Z., Janex A., Monnier B. and Rey D. (1977). Coordinate decentralized control of a complex distillation pilot plant.
IFAC Digital Computer Applications to Process Control 1977.

Brice R.S. and Browne J.C. (1978). Feedback coupled resource allocation policies in the multiprogramming multiprocessor computer system.
CACM Vol.21, No.8, August 1978.

Brinch Hansen P. (1970). The nucleus of a multiprogramming system.
CACM Vol.13, No.4, April 1970.

Brinch Hansen P. (1977). Distributed processes, a concurrent programming concept.
CACM Vol.21, No.11, 1978.

Caines P.E. and Printis R.S. (1978). The stabilization of digraphs of variable parameter systems.
IEEE Trans. Automatic Control, Vol.23, No.2, April 1978.

Casaglia G.F. (1976). Distributed computing systems, a biased review.
Euromicro, Vol.2, No.4, October 1976.

Casey L. and Shelness N. (1977). A domain structure for distributed computer systems.
University of Edinburgh, Internal report CSR-8-77.

Chou W. (1978). Integrated optimization of distributed processing networks.
AFIPS.NCC Vol.47, 1978.

Conant R.C. (1972). Detecting subsystems of a complex system.
IEEE Trans. Systems, Man and Cybernetics, September 1972.

Corfmat J.P. and Morse A.S. (1976). Decentralized control of linear multivariable systems.
Automatica, Vol.12, pp.479-495, 1976.

Desautels E.J. (1978). The many faces of distributed computing. Computing Systems for real time applications.
7th Texas Conference on Computing Systems, Houston, November 1978.

Dijkstra E.W. (1975). Guarded commands, non-determinancy and formal derivation of programs.
CACM Vol.18, No.8, August 1975.

Dijkstra E.W. (1968). The structure of the T.H.E. multiprogramming system.
CACM Vol.11, No.5, May 1968.

Dufour J. and Gilles G. (1978). A methodo-logy for structural analysis and partition of dynamic complex systems. Application to two macroeconomic systems.
IFAC 1978 World Conference.

Enslow P.H. (1978). What is a "distributed" data processing system?
Computer, January 1978.

Farber D.J. and Larson K.C. (1972a). The structure of a distributed computing system software.
1972 Symposium on Computer Communications Networks and Teletraffic, Polytechnic Institute of Brooklyn.

Farber D.J. and Larson K.C. (1972b). The system architecture of the distributed computer system. The communications system.
1972 Symposium on Computer Communications Network and Teletraffic, Polytechnic Institute of Brooklyn.

Fernández E.B. and Bussell B. (1973). Bounds on the number of processors and time for multiprocessors optimal schedules.
IEEE Trans.Comput. Vol. C-22, No.8, pp. 745-751, August 1973.

Firestone R.G. (1977). An analytic model for parallel computation.
AFIPS NCC Vol.46, 1977.

Flon L. and Suzuki N. (1978). Consistent and complete proof rules for the total correctness of parallel programs.
19th Symposium on Foundations of Computer Science, Ann Arbor, Michigan, 1978.

Gaulding S.N. and Lawson J.D. (1976). Process design engineering. A methodology for real time software development.
2nd International Conference on Software Engineering, San Francisco, California, 1976.

Hassan M. and Singh M.G. (1976). The optimization of nonlinear systems using a new two level method.
Automatica Vol.12, pp. 359-363, 1976.

Ho Y.C. and Chu K.C. (1972). Team decision theory and information structures in optimal control problems. Part 1.
IEEE Trans. Automatic Control Vol.17, No.1. February 1972.

Ho Y.C. and Chu K.C. (1973). On the equivalence of information structures in static and dynamic teams.
IEEE Trans.Automatic Control Vol.18, No.2, April 1973.

Ho Y.C. and Chu K.C. (1974). Information structure in dynamic multi-person control problems.
Automatica Vol.10, pp.341-351.

Ho Y.C., Kastner M.P. and Wond E. (1978). Teams, signalling and information theory.
IEEE Trans.Automatic Control Vol.23, No.2, April 1978.

Hoare C.A.R. (1978). Communicating sequential processes.
CACM Vol.21, No.8, August 1978.

Infotech State of the Art Report (1976). Distributed Systems.
Infotech International 1976.

Infotech State of the Art Report (1977). Distributed processing.
Infotech International 1977.

Jain R.K. (1978). Control theoretic approach to computer systems performance improvement.
Proc. of the 14th Computer Performance Evaluation Users Group Conference. CPEUG. NBS special publication 500-41, 1978.

Jones A.K., Chansler R.J., Durham I., Feiler P. and K. Schwans (1977). Software management of Cm*. A distributed multi-processor.
AFIPS NCC Vol.46, 1977.

Keller R.M. (1976). Formal verification of parallel programs.
CACM Vol.19, No.7, pp.371-384, July 1976.

Kimbleton S.R. and Schneider G.M. (1975). Computer communication networks: approaches, objectives and performance considerations.
Computing Surveys Vol.7, No.3, September 1975.

Kobayashi H., Hanafusa H. and Yoshikawa T. (1978). Controllability under decentralized information structure.
IEEE Trans. Automatic Control Vol. AC-23, No.2, April 1978.

Kramer J. and Cunningham R.J. (1978). Towards a notation for the functional design of distributed processing systems.
IEEE International Conference on Parallel Processing, pp.69-76, 1978.

Lamport L. (1977). Proving the correctness of multiprocess programs.
IEEE Trans. Software Eng. Vol. SE-3, No.2, March 1977.

Linton T.D., Tacker E.C., Sanders C.W. and Wang T.C, (1976). Computational and performance aspects of static decentralized controllers.
IEEE Decision and Control Conference 1976.

Lister A.M. (1977). Inter process communication mechanisms.
University of Queensland, Dept. of Computer Science, Technical Report. 1977. St. Lucia, Queensland, Australia.

Lucas M. (1975). Primitives de synchronisation pour langages de Haut Niveau. Seminaire de Programmation.
Laboratoire d'informatique Université Scientifique et Médicale de Grenoble, France.

Maciejowski J.M. (1978). Model discrimination using an algorithmic information criterion.
University of Cambridge. CUED/F-CAMBS/TR174, 1978.

McCabe T.J. (1976). A complexity measure.
IEEE Trans. Software Eng. Vol.SE-2, No.4, December 1976.

Meditch J.S. (1978). Multivariable control of data networks.
IFAC 1978 World Conference.

Mills D.L. (1976). An overview of the distributed computer network.
AFIPS NCC Vol.45, 1976.

Moss F.H. and Segall A. (1976). Progress in the application of optimal control theory to dynamic routing in data communication networks.
IEEE Decision and Control Conference 1976.

Nader A. (1979a). Distributed Computer Control Systems. An Overview.
Dept. of Computing and Control, Imperial College of Science and Technology, Research Report 79/24, London.

Nader A. (1979b). Petri Nets for real time control algorithms decomposition.
IFAC 1979. Workshop on distributed computer control systems, Tampa, Florida, 1979.

Paige R.M. (1977). On partitioning program graphs.
IEEE Trans. Software Eng. Vol.SE-3, No.6, November 1977.

Parnas D.L. (1972a). On the criteria to be used in decomposing systems into modules.
CACM. Vol.15, No. 12. December 1972.

Parnas D.L. (1972b). Information distribution aspects of design methodology.
Proc. of IFIP Congress on Information Processing. Ljublyana. Yugoslavia.

Petrov B.N., Pugahev V.S., Ulanov G.M., Krinetsky O. Ye and Ul'yanov S.V. (1978). Informational foundations of qualitative theory of control systems.
IFAC 1978 World Conference.

Ravi C.V. (1976). The structure and characteristics of distributed systems.
2nd International Conference on Software Engineering. San Francisco California. 1976.

Ross D.T. and Schoman K. Jr. (1977). Structured analysis for requirements definition.
IEEE Trans. Software Eng. Vol. SE-3 No. 1. January 1977.

Ross D.T. (1977). Structured analysis (SA): A language for communicating ideas.
IEEE Trans. Software Eng. Vol. SE-3 No.1. January 1977.

Salter K.G. (1976). A methodology for decomposing system requirements into data processing requirements.
2nd International Conference on Software Engineering. San Francisco, California. 1976.

Sandell N.R., Varaiya P., Athans M. and Safonov M.G. (1978). Survey of decentralised control methods for large scale systems.
IEEE Trans. Automatic Control. Vol. AC-23, No. 2. April 1978.

Sander C.W., Tacker E.C., Linton T.D. and Ling Y.S. (1976). Design and evaluation of large scale state estimation algorithms having information exchange.
IEEE Decision and Control Conference 1976.

Sander C.W., Linton T.D. and Tacker E.C. (1978) Information exchange in decentralized filters via interaction estimates.
IFAC 1978 World Conference.

Schoeffler J.D. (1978). Software architecture for distributed data acquisition and control systems.
IFAC 1978 World Conference.

Schoute F.C. (1978). Decentralized control in packet switched satellite communications.
IEEE Trans. Automatic Control. Vol. AC-23, No. 2. April 1978.

Siljack D.D. (1978). On decentralized control of large scale systems.
IFAC 1978 World Conference.

Singh M.G. and Hassan M. (1977). A two level prediction algorithm for nonlinear systems.
Automatica. Vol. 13 pp. 95-96. 1977.

Singh M.G. and Hassan M. (1978). Hierarchical optimization for nonlinear dynamical systems with non separable cost functions.
Automatica. Vol. 14 pp. 99-101. 1978.

Singh M.G. and Titli A. (1978). Practical hierarchical optimization and control algorithms.
IFAC 1978 World Conference.

Solomon M.H. and Finkel R.A. (1978). A multi-microcomputer operating system.
2nd Rocky Mountains Symposium on Microcomputers. 1978.

Spang III H.A. (1976). Distributed computer systems for control.
SOCOCO 1976. 1st IFAC/IFIP Symposium on software for computer control. 1976

Stone H.S. (1977). Multiprocessor Scheduling with the aid of network flow algorithms.
IEEE Trans. Software Eng. Vol. SE-3, No. 1. January 1977.

Stone H.S. (1978). Critical load factors in two processor distributed systems.
IEEE Trans. Software Eng. Vol SE-4, No. 3. May 1978.

Sundaresham J.K. (1976). Decentralized observation in large scale systems.
IEEE Decision and Control Conference. 1976.

Szlanko J. (1977). Petri Nets for proving some correctness properties of parallel programs.
IFAC/IFIP Workshop on real time programming. Eindhoven 1977.

Teichroew D. and Hershey E.A. III. (1977). PSL/PSA A computer aided technique for structured documentation and analysis of information processing systems.
IEEE Trans. Software Eng. Vol. SE-3, No. 1. January 1977.

Trakhtengerts E.A. and Berezovsky B.A. (1976). Request rank determination with respect to a vector criterion in multiprocessor systems.
SOCOCO 1976. 1st IFAC/IFIP Symposium on software for computer control.

Wang S.H. and Davison E.J. (1978). Minimization of transmission cost in decentralized control systems.
International Journal of Control. Vol. 28, No. 6. 1978.

Williams T.J. (1978). Hierarchical control for large scale systems - a survey.
IFAC 1978 World Conference.

Wulf W.A., Cohen E., Corwin W., Jones A., Levin R., Pierson C. and Pollack F. (1974). Hydra: The kernel of a multiprocessor operating system.
CACM. Vol. 17, No. 10. October 1974.

DISCUSSION

Steusloff: I would like to make a comment about the idea of "observation." Observation here is used for modeling the plant, but it is also a very useful tool for coping with failures in measurement devices and so on. In exception handling in process computers, there can be a lot of the plant variables that can be observed and, if the measuring devices go down, you can switch over to an observer and replace the plant variable you can't measure with the observer output.

AlDabass: Your comment implies that we are using the observer to identify the system. That's not really true. The observer, in this case, is just trying to reconstruct the state of the system. Although, in the larger sense, the observer could act as an identifier of the parameters as well. You end up with a nonlinear observation in that case. By replacing the parameters by state variables and augmenting the state vector, you end up with a much larger state vector, the first components of which are the old state variables and the last part of which are the parameters that you are trying to estimate and identify.

Lalive: I refer to the Figure 2 and I'm asking myself if it's really true that one of the inputs to the software design is the distributed real time operating system. If that is really true, I would really doubt this whole picture because I think that is exactly what we want to avoid. The output of this whole design is computer architecture and we want to design our software independent of any sort of real time operating system.

AlDabass: Well, I think it's the way the diagram is drawn. I wouldn't say that the only output of the design is the computer architecture. You have multiple outputs. The first output that you get is the control functions; you map these onto software modules and then you map these onto the hardware that is going to do the actual work for you. So you have, I suppose, three layers of output, if you like. I suppose the output that we really see is the hardware since that's the thing that occupies the physical space, so to speak. This is probably what is meant here.

Lalive: Of course, as long as the figures are not more specific, one can always argue whether the input should be a level higher or lower. I only want to make mention that it is very dangerous to make this sort of figure.

The other comment would be that at a certain level Nader introduces verification. I think you have to include verification at every level because, if you only start to do verification at some later level, it's not possible if you didn't think of that from the very beginning.

AlDabass: Yes, I think he's using the word verification in the conventional sense of associating it with software design. Obviously, as you said, verification is really inherent in all the levels, starting right from the model design.

Sandmayr: But verification, in the sense of software design, should also start with the software decomposition of the design itself, which you have shown two or three levels above. You can't put it at the end of software implementation; it's part of the software design.

AlDabass: Yes, it's really a matter of how much detail you prefer to put in a diagram of this size. I mean let's be fair; it's trying to squeeze the whole problem of designing DCCS on one page which is virtually impossible to do, and dangerous as well!

Humphrey: Just a comment about this diagram. This is a good review paper but I think that one of the things to consider for a subsequent paper is how you put feedback loops into such a system, and what are the measures of efficiency and effectiveness at some point. I really believe that control systems and plants are living breathing things that change with time. Every two or three years, or whatever the time constant is, changes are going to occur in the system and that has to impact how we do system design. I think it's an area that we are going to have to be addressing more and more as we get into this field.

AlDabass: I would like to make just an inquiry. You mentioned feedback. Do you mean feedback in the classical sense that exists at the controller level?

Humphrey: I mean it in the very sloppy sense of design procedure.

Yoshii: You defined the control of a plant as a set of control functions. How do you define the interaction between control functions? I understand solving interactions is the most important thing in DCCS.

AlDabass: Each function would have a set set of input variables and output variables. The input variables to each function would not only be its own output variables or state, but also the output variables from the other functions. You define each function as having to be fed from the output. For example, take function one; that has to be fed from the output of function 3 and 5, and it's a PID controller of some sort.

Yoshii: In that case, the control function must not be a mere set. There must be some relation between those control functions.

AlDabass: Well, yes, it's not a set of constants, if that's what you mean. It's a set of control functions.

Yoshii: On page 2, you define IP as a set of three sets, where IP is an "information pattern." All of a sudden, that kind of definition enters. This must be derived from the first definition of control functions.

AlDabass: Nader is defining the information pattern as having three entries: There is the implicit information, which is what is embedded into the design decisions, as Nader calls them; there is the structural information, which he calls nonchanging information while the system is in operation, and these are typically the A, B and C matrices; the third entity is functional information and these are usually the state of control variables.

Yoshii: This must represent the interaction that I mentioned.

AlDabass: Yes, the interaction comes in at any one of these stages. It could be at the implicit information level, it could be at the structural information level, or it could be at the functional information level.

Yoshii: So this must be defined on the first definition of a control function which is decomposed into the F1 to Ff. So IP must be defined on the basis of a predefined relationship between control functions. I'm afraid you are missing the definition of the interactions between control functions. You need information exchange between subsystems because you have interactions between control functions. So, in that sense, IP should be defined on some definition of interactions at the level of control functions.

AlDabass: Yes, well that's right. I won't argue with that. But, you see, it's defined at these three levels. You have the control functions, which you then map into software modules, and then you assign the subsystem modules into computation locations. The information pattern has embedded in it all these three aspects. Information exchange takes place at the control function level, it takes place at the software module level, and then finally it takes place at the computational location level. So the information pattern is implicit in all these three levels.

Harrison: I think we should give David (AlDabass) our special thanks. Presenting someone else's paper is difficult, and David has done an excellent job of representing Dr. Nader's ideas.

DEVELOPMENT OF DISTRIBUTED CONTROL SYSTEMS

F. J. Romeu

Honeywell Inc., Fort Washington, Pa., U.S.A.

Abstract. The development of Distributed Control Systems requires management teams integrated by Marketing, Design and Development Engineering and Manufacturing Engineering Departments. These teams plan, organize, and control company resources to provide users with cost-effective systems, more reliable and useful than centralized integrated configuration.

Marketing Engineers are responsible for product management, which includes product planning, functional specifications, cost monitoring, evaluation of sales forecasts, and product support. Subjects to be resolved include product warranties, spare parts, service and maintenance, training of vendor and user personnel, marketing plans, and project management of systems sold.

Design and Development Engineers are responsible for applying proven technology and investigating new technology trends. Engineering designs the product to marketing specifications maintaining a degree of autonomy in matters concerning system architecture, data base structure, communication protocol, software development, components selection, logic design, and testing of engineering prototypes.

Manufacturing responsibility includes production planning, inventory control, machine shop loading, product quality, testing, and directing material purchasing.

Because Distributed Control Systems are intended to be manufactured, installed, and serviced worldwide, the product development strategies become even more complex.

The objective of this paper is to provide the reader with an overall understanding of the considerations that must be taken into account in the development of a Distributed Control System.

INTRODUCTION

Distributed Control Systems provide technological innovation of electronic instrumentation in the field of Process Measurement and Control. These systems are based on microprocessor technology that offers increased control system performance, reliability, and adaptability to process control requirements.

The development of Distributed Control Systems establishes an important and unique vendor company image in the process control industry, broadening the company markets where its instruments and computers are utilized.

International coordination is most important because most of the activities associated with the development, manufacturing, marketing, and sales of distributed control systems are international in nature. A major emphasis is therefore placed on securing worldwide agreement on policies for the selling and implementation of these systems.

The multinational nature of the product requires identification of worldwide marketing requirements, foreign government commercial policies, overseas manufacturing capabilities with convenient system assembly and test locations.

In order to support this product in different countries, the vendor company resources have to be reassessed in the areas of project management, training, documentation, spare parts, customer site support and maintenance services. Fig. 1 shows a general customer support organization.

THE DISTRIBUTED CONTROL SYSTEM CONCEPT

A basic distributed control system, as shown in Fig. 2, incorporates the latest technological developments in digital control systems. Its architecture integrates microprocessor controllers interconnected by a redundant communication link that permits centralization of plant operations. The operator-process interface is based on cathode ray tube (CRT) operator stations. The system function can be expanded to provide computer supervisory or direct digital control, plant optimization, and management information systems.

THE MARKETING PLAN

The Marketing Plan is the document that guides the development of Distributed Control Systems. It addresses the following areas:

- Key marketing objectives and a timetable for their completion
- Worldwide market definition and appraisals
- International subsidiaries' sales forecasts
- Product development strategies
- Product introduction activities
- Field sales program
- Worldwide product support
- Key financial data, including sales revenue forecast, proposed marketing expense, and planned profit
- Pricing strategies
- Field maintenance and service plan
- Advertising and public relations program

PRODUCT DEVELOPMENT DEPARTMENTS

The new product development departments include:

- Product Development Marketing
- Product Development Sales
- Design and Development Engineering
- Manufacturing

The proper staffing of these departments will ensure the timely meeting of the Marketing Plan objectives and coordination requirements.

PRODUCT DEVELOPMENT MARKETING

Product Development Marketing functions include:

- Coordinate worldwide marketing activities concerning product
- Contribute to the formulation of international policies concerning the product
- Evaluate domestic and international regions forecasts covering market size, sales volume, market penetration, and competition
- Present cash flow and return on investment to management to obtain approval for project initiation
- Define product functional requirements utilizing inputs from market managers and international regions
- Formulate performance specifications including environmental requirements, approval bodies, etc.
- Perform competitive analysis
- Determine target product cost
- Recommend appropriate selling price
- Monitor product development costs
- Determine product introduction date and subsequent releases
- Provide support to other organizations

Fig. 3 shows overall activities performed by Product Development Marketing.

PRODUCT DEVELOPMENT SALES

Product Development Sales functions include:

- Plan market penetration (geographic, industry, product)
- Build a skilled team, properly equipped and motivated to reach the company sales goals
- Pursue selected segments of growth markets with high profit potential
- Define and plan product penetration within designated key accounts that meet specific conditions such as:

    - Good relationship with vendor company
    - Influential in their market
    - Strong engineering staff

- Large business potential
- Multinational
- Process computer oriented

. Interest customer in a complete system utilizing advanced technology with capacity to grow with future plant expansions
. Evaluate countries where the product can be sold, using as criteria:

    - Level of customer support that vendor can offer in that country
    - Foreign government commercial policies
    - Cultural influences

. Recommend appropriate pricing strategies to management
. Formulate commissions and incentives policy
. Assist in the integration of customer support functions such as:

    - Project Management
    - Training
    - Documentation
    - Service
    - Spare Parts Availability

. Contribute to the formulation of the official terms and conditions of sale
. Coordinate the advertising program with Marketing Communications
. Support field sales organizations

Fig. 4 shows overall Product Development Sales activities.

Pricing Strategies

Pricing strategies are required to achieve:

. Price stability in the industry
. Forecasted volume of sales
. Target profit margins and desired market share

Price discounts are utilized as a sales resource in the pursuit of strategic projects and to increase or decrease sales in function of manufacturing capacity.

Pricing strategies include spare parts pricing and exchanged parts trade-in allowance.

Commissions and Incentives

Adequate commissions and incentives have to be given to the international subsidiaries to encourage them to find new business and give application engineering support, warranty, training, and third party commissions, if any.

Conflicting sales performance incentives must be avoided.

ADVERTISING

The main purpose of Advertising is to cut selling costs, reducing the cost-to-sell ratio while broadening and/or deepening the awareness of the vendor company capabilities for supplying distributed control systems and services.

There is a need for strong and continuing communications with the customer management and engineering levels.

The selected communications media include:

. Publications space advertising
. Selectively programmed direct mail
. News releases
. Technical papers
. Trade shows
. Sales literature
. Audio-visual presentations support

CUSTOMER SUPPORT FUNCTIONS

Project Management

Project Management must coordinate all activities associated with instrumentation and control systems involving multinational organizations, from purchase order to system start-up.

As shown in Fig. 5, the Project Manager is the focal point of all customer/contractor requirements and the vendor's operating departments.

The foreign subsidiaries, under the Project Manager's direction, provide effective assistance to the process licensors and subcontractors to ensure proper support.

The Project Manager has appropriate staff assisting him in the technical, planning, contract administration, and quality assurance areas.

Project Management responsibilities include:

- Formal project communication
- Defining project specification
- Project scheduling
- System engineering, assembly and test
- Quality assurance
- Training of customer personnel
- Documentation
- Spare parts
- Shipping
- Commissioning and start-up, if required
- Invoicing

Training

Training programs are conducted for in-house personnel, field sales, and customer/contractor staff. Customer training courses are primarily directed to plant operations, service technicians, and project engineers.

Because of the different languages required, training centers are located in different parts of the world. Each one of these centers has a full complement of dedicated training equipment to permit hands-on experience.

Documentation

The product documentation includes:

- Product specifications and technical data
- Operating and installation instructions
- Maintenance and service manuals

Selected documents are provided in different languages.

Each control system delivered to a customer is also provided with system design documentation that includes:

- Hardware layout
- Wiring list
- Power wiring
- Field connections
- Loop identification and interconnection wiring
- Controller configuration

Service

The recommended method for correction of equipment faults is to replace the defective part with the corresponding spare part. The equipment has been designed to reduce to a minimum the need for performing repairs at the job site.

The digital instrumentation performs self-diagnostics that identify many of the components in which the fault has occurred. The corrective maintenance procedure is to replace the identified component. Fault analysis and repair are performed only at the manufacturing location.

Normal preventive maintenance items that require attention are:

- Cleaning of air filters, CRT screens, etc.
- Periodic checks and calibration of:
    - Analog displays
    - Recorders
    - Auxiliaries
    - A/D converters
    - Regulator cards
    - Output cards
- Periodic testing of back-up systems such as:
    - Redundant communication channel
    - Uninterrupted automatic control
    - Battery back-up

Spare Parts

Spare parts are located at the following locations:

- The customer site
- Within the service organization
- The manufacturing location

Spare parts at these locations ensure a successful repair/exchange program. Spare parts, located within the service organization in support of the customer system, will be ordered, shipped from the factory, and placed in the depot well in advance of the actual system installation.

There are regular reviews in spare parts inventory control. This assures that inventory levels and factory orders are adjusted to meet field requirements.

DEVELOPMENT ENGINEERING

Development Engineering is responsible for the design and development of:

- Electromechanical recording, indicating, and controlling instruments
- Field transmitters, control valves, process analyzers
- Distributed control systems and computers

This organization must make sure that products do not infringe on any U.S. patent covering the design of the equipment.

Engineering designs products to allow their manufacture at minimum cost at any worldwide manufacturing location, without modification of the design.

Devices and parts having the same model or part number must meet the same performance, appearance, and customer interface specifications; and must be interchangeable in all respects, regardless of where they are manufactured.

Any model number and options added to the product line in any worldwide manufacturing location must have worldwide marketing approval.

The functional criteria of distributed control systems are oriented towards an overall reduction in system installation and field equipment wiring costs, emphasizing system tolerance and security against faults caused by either equipment failure or human error.

With these criteria in mind, Engineering determines the system architecture, data base structure, communication protocol, software/firmware development, component selection, logic design, and testing of product prototypes.

Fig. 6 shows the Development Engineering department interfaces.

Product Design Control Documentation

The product design control documentation consists of:

- Model Specification (MS), to define the model number designations which are used to document products in the line
- Product Specifications, to define the performance characteristics of all elements of the product line
- Drawings and engineering specifications
- All documents required to identify and assemble the product

PRODUCT ASSURANCE

Product Assurance's mission is to assure customer satisfaction with the product.

Functions performed by the Product Assurance organization include:

- Evaluate new product design and competitive products
- Plan and inspect new product quality
- Develop software/firmware for automatic testing systems utilizing mini and micro-computers
- Design and build new product test equipment
- Inspect purchased material, sub-assemblies, final assemblies, and integrated systems
- Monitor worldwide field performance of all products
- Initiate corrective action to improve products

MANUFACTURING

Manufacturing produces the new product in compliance with all applicable requirements of the Fair Labor Standards Act, and regulations of the United States Department of Labor.

Manufacturing Operations, shown in Fig. 7, include:

Tooling. Utilizing the latest auto-sequencing and auto-insertion equipment.

Production Control. Minimizing inventory while maximizing labor utilization. The marketing forecast is broken down by:

- Model Specification (MS)
- Parts list
- Depot usage
- Miscellaneous usage
- Foreign factories

This information is fed into a computerized system that performs all necessary data processing for routing information or bill of materials requirements.

Master Scheduling. Allocates inventory to a customer order to meet its delivery schedule, and enters it in the manufacturing backlog.

Inventory Control. Assigns money for investment in inventory and determines when and how much to purchase.

Shop Floor Control. Applies labor and machines to production, and directs the Purchasing Department by specifying the need for additional material.

CONCLUSION

The development of Distributed Control Systems, from the time of their creation as a concept, to the time when the control systems are operational at the users' plants, represents one of the most challenging accomplishments in the Industrial Control business.

The market-oriented company puts all of its resources to add value to its product and help create sales. There is no dominant department - not marketing, not finance, not engineering, not manufacturing.

The purpose of this paper is to show the different disciplines involved in the development of a Distributed Control System. The coordination among them is built in a logical step-by-step sequence, concentrating every effort on providing the highest possible benefit to the user.

REFERENCES

Williamson, W.E, (1976). The Special Promise of Micro-Computers as Elements in Distributed Computer Control Systems. Honeywell Inc.

Dallimonti, R., (1972). Future Operator Consoles for Improved Decision-Making and Safety. Instrumentation Technology August 1972.

Romeu, F.J., (1979). Reliability of Distributed Measurement and Control Systems. Instrumentation Technology, May 1979.

Romeu, F.J., (1977). "Funcionamiento y Capacidad de Sistemas de Control Distribuido". ISA Seminar Mexico City October 1977.

Romeu, F.J., (1977). Technology of the TDC 2000 Digital Control System. U.S. Department of Commerce Exhibition Instru-Quipos 1977, Caracas, Venezuela November 1977.

Romeu, F.J., (1977). Evaluation of Distributed Process Control Instrumentation Systems. ISA Conference, Anaheim, California, May 1977.

Romeu, F.J., (1976). A Digital Communication System for Process Control Instrumentation. ISA Conference, Houston, Texas, October 1976.

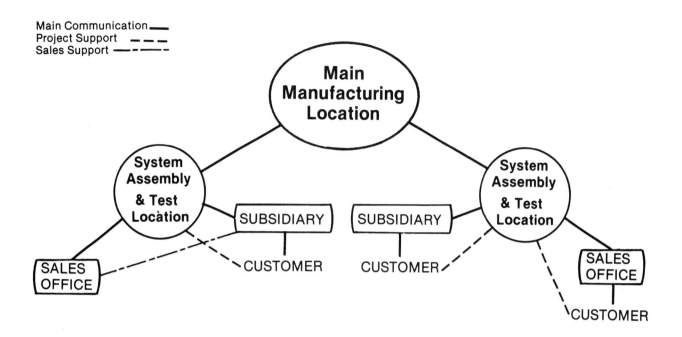

Fig. 1 General Customer Support Organization

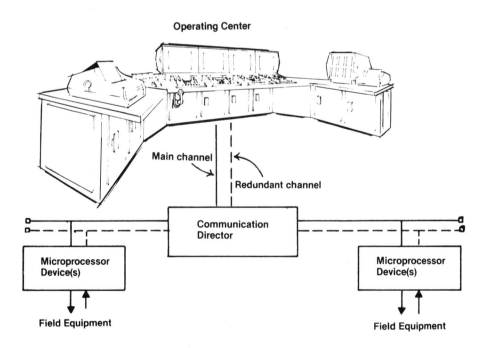

Fig. 2 Basic Distributed Control System

Fig. 3   Overall Product Development Marketing Activities

Fig. 4   Overall Product Development Sales Activities

Development of Distributed Control Systems 191

Fig. 5 Project Manager Interfaces

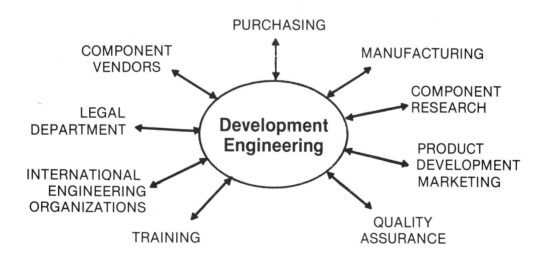

Fig. 6 Development Engineering Interfaces

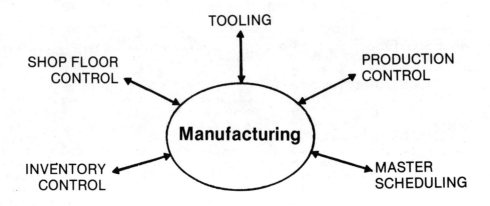

Fig. 7   Manufacturing Activities

DISCUSSION

Harrison: Before starting the questioning, I would like to say something about why I accepted this paper for the workshop. It is, perhaps, an unusual paper to be presented at what we all realize has been a very technically oriented workshop conference, but I accepted it for a very specific reason. Very often, and in fact it's happened here, you hear questions such as why don't you vendors build equipment this way, or have you ever thought of doing this, or why can't I get that spare part instantaneously when I need it? I thought it might be very interesting to explore these questions and to provide a chance for users of computerized equipment to hear the story from the vendor side of the table. It's very clear from Frank's statement that the vendors are in business in the same sense the users are in business; that is, to provide a return to the stockholder. The story he told could equally be applied to developing automobiles, consumer products or anything else. So Frank has satisfied the goal I had for this paper.

This is an opportunity for those of us at the workshop to have a real live vendor standing up and talking about the way the computer business is run. I would suggest that our questioning should explore those aspects, as opposed to the purely technical aspects of what is underneath the covers.

Humphrey: I have a couple of questions along the lines of what Tom (Harrison) suggests. Could you amplify a little bit more about the market survey? As a user in a national laboratory environment, I've never been aware of being contacted by a national vendor as to my particular application. I would like to hear a little bit more about how that is done.

Secondly, as I was listening to you describe the operation, I was thinking about the Ada project, which has a lot of very similar things, but there was one aspect of it that was very obvious in the Ada project and less obvious in your talk; namely, the fact that there was a lot of opportunity for feedback and correction from the user community in the Ada project. Is there a mechanism for such feedback in commercial projects, taking into account that there is a proprietary problem in revealing the ideas that you plan to market?

Romeu: I am not able to comment on the Ada project, but let me try to answer your first question. Basically, as I understand it, the vendor companies decide to penetrate the market where they feel they have the greatest expertise. They establish a very good communication with the leading user companies in those markets. They seek, and appreciate very much, their input concerning all areas in the development of the project. In fact, in many instances it becomes very much a joint development with these leading companies that are the foremost or most important users of the product. Also, as time progresses, the different customers that buy the systems provide feedback. So in your particular case, when the vendor is directly applying his product to your particular needs, you would be consulted on your opinion. Certainly, it would be most appreciated because, without it, it is impossible to develop any system.

Humphrey: What about the second question? I didn't really care about a discussion of Ada itself but I was wondering if there are opportunities in the development cycle, before you start selling, for user feedback on product design?

Romeu: Yes. I can mention to you, as a matter of fact, that a development of a control system requires a number of years. These years are really filled with a lot of meetings, discussions, conversation and correspondence between users and the vendor. The system that we developed took no less than four years to develop and that is without even taking into account the early planning done by a gentleman sitting at this table Mr. Dallimonti. So it just shows you that, certainly, that is the way that systems are done.

Dallimonti: I'd like to add to Frank's answer. In the development of the product that Frank's been talking about, the way we tried to get customer feedback was by getting them to try to work with us. It isn't easy to do. For one thing, there are always a lot of proprietary barriers that are difficult to work through; for another thing, if you are really after an innovative product, it's difficult to get people to talk about it in innovative terms. In fact, I would say that, if you analyzed the feedback we got when we were proposing this product, you wouldn't have marketed it because the vast majority were afraid to endorse such a thing. So it's a delicate and interesting process of trying to work with the user in order to understand his problem, but not necessarily always to accept his solution. That's the real challenge in this phase. So research of this nature is not a polling of opinions and merely taking statistical averages and things of this sort.

Willard: Let me put on my vendor's hat for a moment to answer, from my viewpoint, the question that Rusty Humphrey asked. One of the most promising ways for vendors to do market research is to come to meetings such as this to explore and exchange ideas, and listen to customers' complaints about what we did last year, or what our competitors did last year. From this we try to capture the essense of what the needs of the future are.

Another issue. I have a question to ask you, Frank (Romeu) or Renzo (Dallimonti). Could you expand somewhat on what the differences are in the development cycle between the development of a distributed control system versus a centralized control system? Are there any essential differences in the way you go about development because the product is distributed?

Dallimonti: Well, as I review the history of the development, I would say there were some important differences. In a distributed system, you postulate a number of distributed modules and, as a matter of development organization, they end up being the responsibility of different design groups. The need to coordinate those developments so that the system eventually is able to communicate with itself is one of the main tasks, as well as also being able to keep under control the scope and function of each of those modules so that they have this unified capability. Another thing that we found, and this is probably just a matter of history, was that the concept of a distributed system was actually very hard to sell to the computer people; so one of the largest problems we had was getting the committed involvement of the central computer school in trying to make the distributed system work. In the last few years though, there has been so much momentum built up that I don't see that as the problem anymore, but it was in 1969 when we started.

Lalive: I think it's a fact that more and more users, when introducing distributed systems, begin to be competitors since they are building their own small microprocessor controlled devices whereas, some years ago, they would have solved their problems by buying minis. What's the reaction to this and how do you try to help users to remain users and not to build their own electronics?

Romeu: That's a very good question and, in my mind, it establishes the way that modern technology is coming down to help every single application engineer to solve his problem. If I may speak and later on Dali (Dallimonti) will, I'm sure, be able to help us out. My feeling is that the vendor company has to establish a product specification with very broad guidelines and is not possible for a product to satisfy all users and all applications; there just wouldn't be time in the world to do it. So, in the marketing studies, you establish what your product will do and you establish a certain market, a certain number of users. You hope that, with your know-how and your product, you can help those users. There will be a series of users that will not be satisfied with that particular product and, because of the limitation of your resources, you cannot help everybody in their application. The most you can do is provide assistance to those users that feel that their application would be resolved only by a special system that satisfies their particular application requirements. We help them out by providing these users will all the knowhow we can and, I believe, that's as far as we can go. Maybe Dali (Dallimonti) can expand a little bit more on that.

Dallimonti: I think that's a very relevant observation. I think all that it means, as I see it, is that the vendor has to continually appreciate the user's needs, such that his standard product offering has the scope to satisfy enough of them so that the residual, that's not totally satisfied the way you would have liked, is small enough that it doesn't pay you to try to work in that area. I think we find that most users, who could have the capability to handle many of these problems, just don't feel that it is economically worthwhile to work on that difference element, or they will work on the difference element and use the basic standard solution. The only resort we have is to supply a product which has a broad capability. That means continually appreciating your problems, to know them in detail, and to find the standard solutions. It has got to be standard answers and standard products, because that's the only way we can achieve economy of scale and, when you get right down to it, the process control business doesn't have large quantities.

Legrow: Let me address Mr. Dallimonti for a moment and give you a user perspective. It is a fact, as Rick Signor has alluded to, that we end up putting in a smorgasbord of equipment from various vendors. As much as IBM or Modcomp or Honeywell would like to think that they are going to capture an entire company's business, it just doesn't happen very often. From a user perspective, the thing that is often frustrating is that a general interface to the rest of the world often appears as an afterthought. This is the thing that I would like to address to all of the vendors. Why does it always appear that this is the case, and what can be done about it?

Dallimonti: Well, that is a good point. I can only reflect historically that everytime you try to evolve what you think is a comprehensive solution, there is always the specialist that has a uniquely directed solution focused on that particular problem of yours to try to satisfy it. So, the solution takes a more exact form than your standard variation. The other thing, having to do with the general purpose computer interface, is that one is wary of going in that direction because we become involved, whether we like it or not, with the problems of applying a foreign device to our system. One of the problems of these distributed systems is the need to know that it's remaining integrated and how foreign devices may or may not interact with the protocols and the structures that you have built in. As we have seen, we are all still learning about how to design distributed systems, so, in this specific case, we were concerned about exposing ourselves to the problems of everyone else's computer.

Harrison: If you want another vendor point of view, I would point out that the word "general," which you have used several times, is so imprecise that it is in the source of some of the problems that the vendor has. When you say, let's provide a general interface for all of these devices, the problem is that the devices don't present the computer with an interface which is the same. So, one runs into the problem of precisely what is a general interface. I know that for a lot of companies the solution has been to provide 16 points of DI, 16 points of DO, and a couple of interrupt lines, and then you build your own interface through the software, recognizing that it probably is not optimum. The vendor is faced with more different interfaces than any single user is because, in most cases, the vendor is looking at interfaces for equipment in many different areas, data processing kinds of equipment, versus digital PID controllers, versus RS232 devices. So we, in fact, really have more interfaces to think about than any single user ever would have. It is a very hard problem: What is a flexible interface that will solve enough problems to be a useful and commercially viable feature on a system?

Gellie: I have a related question. How do vendors view conferences and workshops of this nature, and, if they have any value, what specific values do you see in these meetings and, similarly, standards and standard committees?

Romeu: I would like to call on the help of Dr. Harrison who is intimately involved in this matter. He certainly can serve as a spokesman for the vendor industry.

Harrison: I think the answer is obvious. Clearly, if they were not of value most of us wouldn't be here. They are of value in a number of different ways. One is Renzo's (Dallimonti) reference earlier to understanding what users need; these meetings have a great value in that area. Standards committees provide, I think, many of the same advantages and many of the same opportunities. The thing that has to be differentiated between, I believe, is the effort that anyone, a user or vendor, is willing to put into a technical organization or standards committee versus any implied direct action such as automatic agreement to implement a resulting standard. These questions, in fact, are quite distinct. There is a technical question having to do with development of workshop sessions, of technical papers, and of spreading the knowledge versus the business side which is, pure and simple, business. So I think there is value. If there was not value, you wouldn't see the vendor representation at the standards meetings, at the technical society meetings, and that type of thing.

Willard: Let me point out that every engineer, whether he works for a vendor or a user or an OEM, thinks he is a problem solver. What we want to do is to make sure we are solving the right problems.

In regard to the questions about interfaces and standardization of interfaces, let me recall that we gave you an interface once and all you did was beat us over the head with it. I refer to RS232. RS232 took many, many years to settle down; it's now in revision C. It's changed quite a bit and it's wrong, it is technologically backward. We are stuck with it; it causes problems for you and it causes problems for us. I don't really want to see us prematurely standardize on something else that will cause us as much grief as I believe RS232 has.

Legrow: RS232 may have caused us some grief, but I'm sure it has saved us a lot more grief than it has caused us. For many of us, that is the only way we will be able to put in equipment from more than one vendor.

Willard: The specific problem from the vendor's side is that every time somebody connects something via RS232 and encounters a grounding problem, the vendor takes the blame and we get into a lot of trouble, even though the real cause is the 232 requirement to carry the ground through the interface. We don't want to do that again.

Wilhelm: I think that one major difficulty in providing general solutions or general purpose interfaces is that many, many people confuse the concepts of generality and universality. A lot of people tend to look for universal solutions. I think that this applies, not only at the application level, but at other levels in the system where we are talking about communication protocols and so forth. It is very unlikely that one will find a universal communication protocol, a universal interface, a universal controller or whatever, but one can come up with some general tools that can be applied to a number of different applications.

PETRI NETS FOR REAL TIME CONTROL ALGORITHMS DECOMPOSITION

A. Nader

Imperial College of Science and Technology, Computing and Control Department, London SW7, U.K.

Abstract. When a real time control algorithm is to be executed in several computational facilities (multiple processor, multi-processor, network etc.), the problem of how to decompose the control algorithm arises. The decomposition should consider the concurrent process synchronization, interprocess communication and performance problems. In this paper a single formal tool that allows one to treat different aspects of the above problems is developed: Time Valued Petri Nets.

Time Valued Petri Nets are a simple tool for modelling, decomposing and analysing centralized real time control algorithms for parallel processing. A great variety of aspects can be analysed with this tool: decomposition, data flow, concurrency, error propagation, execution time, interprocess communication, trade off of system management requirements and number of processes, assignment of processes to processors and number of processors required.

The application of Time Valued Petri Nets to a life size multivariable model reference adaptive control algorithm has shown the ease and power of its use.

Keywords. Distributed Computer Control Systems, Multiprocessing systems, Parallel Processing, Petri Nets, Communications Computer Applications.

INTRODUCTION

Distributed Computer Control Systems (DCCS) have only recently become feasible. DCCS arise with the necessity to control systems
- that are geographically or functionally distributed
- that impose information communication restrictions
- where the performance and/or reliability of a single computer is inadequate to meet the requirements of the control system.

The advantages and disadvantages of distributed control compared with centralized control are analysed in (Jenkins, 1978). The centralized approach has the transitional but considerable advantage of already existant hardware and software for real time control. The essential disadvantages of the centralized approach are: the cost, including wiring and communication interfaces; cost effectiveness is low; there is a lack of ease of implementation and lack of flexibility.

The advantages of a distributed control are essentially the complement of the centralized approach. Distributed control allows ease of program implementation, down line loading; it relieves communication cost overheads and is cost effective in terms of computational power. Distribution of processing power at the right location in the plant improves response time to the process, while reducing the work load of the host. Distribution enhances flexibility, modular incremental growth and increases reliability. A common justification for hierarchical networking is the high degree of reliability attained by spreading the monitoring and control functions across many smaller computers by providing one or more levels of backup at each node of a network.

Essentially the most cited claims for distributed control are (Schoeffler, 1978), (Syrbe, 1978), (Spang, 1976), (Gaspart, 1978), (Horton and colleagues, 1977):
- high cost effectiveness
- high reliability
- high availability
- high performance
- modularity
- ease of implementation
- graceful degradation

Absolute flexibility or "dynamic restructuring", feature so emphasized in distributed processing literature, is less likely to be the case in DCCS where the computation locations must contain programs relevant to that location either geographically or functionally, or even more, physically constrained to that location because of hardwired connections to sensors and actuators.

At present the major disadvantages of distribution are that the software required to support network communications can be very complex, often forcing system designers to generate their own network support programming (O.S. and communications). Furthermore the problem of distribution; partition of system and application software is not yet completely solved. The most common distribution rule (Spang, 1976) is that a system should be divided into functional parts, in such a way that there exists maximum independence, with well defined simple interfaces and a minimum of requested communications. Other quantitative performance studies show fairly clearly that performance improves if and only if communication between processors is minimized (Raskin, 1978).

Distributed processing for process control in its widest sense is not only restricted to local area networks but also includes, multiple processors; multi-processors, etc. (Nader, 1979). When a centralized control algorithm is to be executed in several computational facilities concurrently (multi computer, multiple processor, multi-processor, etc.), the problem of how to decompose the control algorithm arises. The decomposition should consider the concurrent process synchronization, interprocess communication and performance problems.

In this paper a single formal tool that allows one to treat different aspects of the above problems is presented: Time Valued Petri Nets. A Time Valued Petri Net as defined in this paper is a safe, pure Petri Net; where places have one of two semantic meanings: process execution or data communication. Transitions are interpreted as start or completion of a process execution or data transfer. To each place is associated a completion time.

Standard Petri Nets allow the analysis of concurrency of processes whereas the approach presented in this paper allows in addition:
- the comparison of different decompositions
- data flow analysis
- program verification and error analysis
- estimation of minimum execution time to calculate a control action
- tradeoff analysis of management of the computer system, degree of parallelism and number of processes
- determine the number of processors required for a given implementation
- assignment of processes to processors
- interprocess and interprocessor communication analysis.

Functional decomposition of centralized control algorithms is natural since control blocks have a functional meaning. However different levels of decomposition are possible: the control function level, the mathematical function level, the instruction level, the logical function level, etc. The multiplication of processes in order to obtain maximum parallelism also implies the multiplication of interprocess communication and process synchronization; then the increase of complexity of process management and decrease of performance. The tradeoff of level of decomposition, degree of parallelism and complexity of process management can be analysed with the proposed tool.

The decomposition problem we will deal with in this paper is static (decomposition takes place before execution). Static decomposition means that there exists a fixed binding between processes and processors as opposed to dynamic decomposition where this binding can change during execution of the algorithm (decomposition takes place during execution). The approach we present here allows one to determine the number of processors for a given decomposition and viceversa how to decompose an algorithm to be executed in a given number of processors.

The decomposition of the algorithm will produce a parallel algorithm as a collection of concurrent processes that operate simultaneously for solving the control problem. The decomposition we will treat in this paper produces synchronous parallel algorithms. A synchronized parallel algorithm is a parallel algorithm consisting of processes such that there exists a process that at some stage is not activiated until another process has finished a certain portion of its program (Kung, 1976). The needed timing can be achieved by using various synchronization primitives.

In an asynchronous parallel algorithm there is no explicit dependency, as in synchronized parallel algorithms, between processes. The main characteristic of an asynchronous parallel algorithm is that its processes never wait for inputs at any time, but continue or terminate according to whatever information is currently contained in the global variables. Synchronization is not needed for ensuring that specific inputs are available for processes at any time. This paper will not deal with asynchronous parallel algorithms.

We will first present a general overview of software decomposition. Secondly we present the Time Valued Petri Nets formally and finally an illustrative example is presented. The decomposition of a multivariable model reference adaptive controller is examined using Time Valued Petri Nets. The analysis and comparison of different decompositions of this controller at different levels is presented. This example shows that the decomposition based on control functions is not always the best.

SOFTWARE DECOMPOSITION

Real time process control algorithms

decomposition has to be studied in its more general context of software decomposition. Before distributed processing, software decomposition was needed for the design and analysis of multiprocessing systems. A widely accepted design concept that has been in computing science since that time is that of the "module". The hierarchical design (layered structure) methodology has forced the designers to establish suitable definitions of modules, modularization or decomposition criteria and module interconnection specifications. The need for clear specification of modules and their interconnection has been satisfied following different approaches:
- language approaches
- automata approaches
- graph approaches.

The decomposition problem for distributed systems is formally equivalent to the decomposition problem for multiprocessing. The different approaches that are currently used for multiprocessing are then relevant to the decomposition problem for distributed systems.

Language Approaches

In the language approaches the need for a clear specification of modules and their interconnection has been satisfied on the one hand with the association of "modules" with protection mechanisms and on the other hand reliability and correctness considerations have lead to the introduction of assertions in these modules.

Inherent problems to the "module" conception are the interprocess communication and synchronization. Many different mechanisms have been developed: semaphores, critical sections, monitors, path expressions, distributed processes, communicating sequential processes, etc. The combination of the "module" idea with protection, reliability and correctness concepts together with different interprocess communication mechanisms has produced a wide variety of approaches. (Hansen, 1977), (Hoare, 1978), (Thomesse, 1977), (Kramer and Cunningham, 1978).

The module concept is as old as that of the hierarchical system: hierarchical systems are always formed by well defined entities called modules. The essence of the module concept is its easy semantic interpretation. Logic independence with other modules provides a sound basis for insurance of protection and correctness. The assumptions made about other modules should be minimal, in particular absolutely independant of module implementation (information hiding) (Parnas, 1971). A module can also be considered as a responsibility assignment (Parnas, 1972). Baer (1973) reviews methods for automatic detection of parallelism in high level languages. Language approaches are mostly used at the system design level.

Automata Approaches

The automata approaches conceive software as a coded automaton (Mendelbaum, Madaule 1975). Testing and decomposition (composition) can be done as it is done with automata: test sequences, automata addition, concatenation, etc.

Graph Approaches

Graph approaches are mostly used at the application software level. Graph approaches are less rich than the language ones although used as an analytical tool, they are more efficient with the possibility of quantification of some aspects of the design. Traditionally graph approaches have been the most used for flow analysis. Baer (1973) reviews different models for parallel computation: graph models, Petri Nets, parallel flowcharts, and flowgraph schemata. Data flow and Control flow analysis for computer programs are used more and more. (Hecht, 1977; Kodres, 1978; Allen and Cocke, 1976; Brantley and others, 1977).

The predominant approaches used for partitioning program graphs are (Paige, 1977):
- segments
- DD paths
- intervals
- classes
- level i paths.

A segment is a block of consecutively executable statements with one entry and one exit. (S) nodes can be junction nodes, decision nodes, entry and exit nodes. A segment is then a simple path between two (S) nodes such that no (S) node appear within the segment.

DD path (Decision to Decision path) is a logical program block i.e. a segment of the program that is executed as a single decisional node. (D) nodes are the decision nodes, the entry and exit nodes. A DD path is a simple path between two D nodes such that no D node appear within it.

Intervals are the partition element used for control and data flow analysis. A data flowgraph describes functionally what happens to the data during one execution sequence. Given a node h, an interval with header h denoted I(h) is the maximal single entry subgraph for which h is the entry node and in which all loops contain h. Several algorithms for partitioning a flowgraph into intervals are presented in (Hecht, 1977).

Classes are based on regular expressions. A regular expression can be used to describe the behaviour of a one input one output graph. The graph is partitioned into classes. A class is a sequence of terms grouped under a star (*, + Kleene notation) or combined with another sequence by a cup. Classes are used to group commonly originated graph parts together.

A partitioning concept that is related to classes is that of level i paths. A level o path is a simple path from the input vertex to the output vertex. A level i path is a simple path or loop that begins and ends on nodes that have been elements on paths of lower levels, but all other edges and nodes of the path have not been used on any paths of lower levels.

Given a certain program partitioning for multiprocessing, the problem of optimal scheduling and number of processors can also be analysed through graph approaches (Stone 1977, 1978). Fernandez and Bussell (1973) present a formulation for lower bounds on the minimum number of processors and on time for multiprocessing.

TIME VALUED PETRI NETS

Petri Nets are a powerful modelling tool for concurrent processes. They have already been used extensively in many aspects of software design and analysis (e.g. realtime operating systems). An introduction to Petri Nets is presented in Appendix 1, the reader unfamiliar with this tool should read it before this section.

A Petri Net is a quadruple $N=(S,T,Z,Q)$ where S and T are the two types of nodes of the net (state and transition nodes). Z and Q are the forward incidence function and backward incidence function respectively. A Time Valued Petri Net as defined in this paper associates an estimated time value with the states. The approach we present here is essentially based on the idea of considering data and control flowing together. This is done considering the following process definition:

A system Σ is defined as a collection of processes P_i and their interrelation specification. Using a Petri Net this interrelation specification is given by Z and Q. The system is then:
$$\Sigma = \bigcup_{i=1}^{n} P_i \cup \{Z, Q\}$$
A process P is a quintuplet $P=\{I,O,CX_0,CX,A\}$ where
I are the process inputs.
O are the process outputs.
CX_0 is the initial context (state) of the process.
CX is the context (state) of the process.
A is the process algorithm i.e. the specification $A: I \rightarrow O$.

We particularize this process definition by considering the inputs and outputs not only as data but also as control information.

The interpretation that we make of a Petri Net is considering the states as processes executions or data transfers.

Note that the presence of tokens in the places represents the state of the system Σ and not of the processes.

Definition

A Time Valued Petri Net is a safe, pure Petri Net in which we assign to each place a time value: to processes places an execution time t_e and to transfer data places a data transmission time t_t.

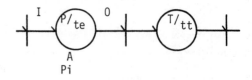

Fig. 1. Time Valued Petri Net

As we are dealing with real time algorithms the Time Valued Petri Net associated with it contains a transition that is at the origin of the net that is fired every period of the real time clock (entry point).

Although we do not rule out the possibility of 'OR' nodes existing in the net, it is unlikely that they appear. OR nodes of the attribution type are:

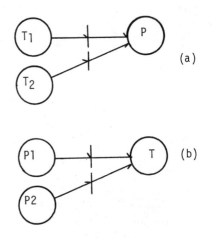

Fig. 2. OR attribution Nodes

Case a) is definitely not possible because a Process P has a well defined input vector. If two input vectors arrive at the process place it means they are the same and then no OR nodes with a) interpretation will appear in this approach.

In the case b) this net is interpreted as two output vectors transferred using the same data transfer procedure. This case is likely to happen but it can be expressed more clearly as:

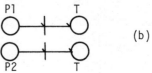

OR nodes of the distribution type are:

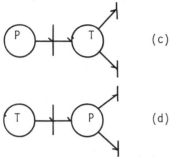

Fig. 3. OR Distribution Nodes

These two interpretations are also ruled out because the output transitions of the process place d) (or of the transmission place c)) will be fired after completion of the process (or data transfer) at the same time producing an ambiguity (hazard). The output transition associated to a place is unique because it corresponds to the completion of the process execution or data transfer that the place represents.

In general although OR nodes of the b) type can occur in a Time Valued Petri Net of a real time algorithm, this can be transformed to a Time Valued Petri Net without OR nodes.

We have seen that a net of this type has a transition corresponding to the real time clock event. Considering this transition as a reference one can calculate the minimal time at which each transition will be fired.

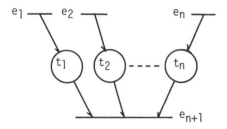

Fig. 4. Minimal firing time

If e_i is the time when transition i is fired then

$$e_{n+1} = \text{Max}\{(e_1 + t_1), (e_2 + t_2), \ldots, (e_n + t_n)\}$$

The firing of the transitions can be made using time dependent events (after a predetermined time has elapsed) or time independant events using correctness criteria events (post or preconditions for completion or start of a process execution or data transfer).

Real time control algorithms modelled with Time Valued Petri Nets constrain these nets to be safe. The presence of more than one token in a place would indicate failure of the algorithm. No OR nodes, one entry point and one end point oblige that all processes and data transfers must be executed and completed before the next firing of the real time clock transition.

Simple Example

To clarify these ideas let us present a simple example. Suppose a real time control algorithm of the type

$$U^o(k) = U_L(k) + U_a(k)$$

A model using Time Valued Petri Nets for this algorithm is:

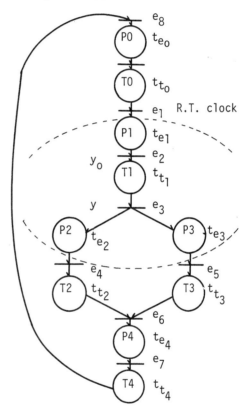

Fig. 5.

where
Po dummy process
P1 sampling plant outputs y_0
P2 computation of U_L
P3 computation of U_a
P4 computation of $U_L + U_a$
T_i data transfer of process Pi
tei execution time of process Pi
t_{t_i} transmission time of data transfer Ti
   t_{t_4} application of the control to the plant.
   t_{t_1} transfer of the sampled plant outputs.

Considering a time dependent event firing the minimum execution time of the algorithm will be e_8.

$e_8 = e_7 + tt_4$
$e_7 = e_6 + t_{e4}$
$e_6 = \text{Max}\{(e_4 + t_{t2}), (e_5 + t_{t3})\}$
$e_5 = e_3 + t_{e3}$
$e_4 = e_4 + t_{e2}$
$e_3 = e_2 + t_{t1}$
$e_2 = e_1 + t_{e1}$
$e_1 = 0$

$e_8 = t_{e4} + tt_4 + t_{e1} + t_{t1} + \text{Max}\{(t_{e2} + t_{t2}), (t_{e3} + t_{t3})\}$

The increasing of processes Pi in order to obtain a maximum parallelism also implies the multiplication of data transfers, then the increase of communication and synchronization management requirements (this expressed in data transfer time). The tradeoff of the level of decomposition, degree of parallelism and management complexity can be studied through this approach; the finer the decomposition, the bigger the number of processes; the bigger the number of processes the briefer the execution times for each process.

Suppose in our simple example a purely sequential execution. The minimum execution time for a purely sequential algorithm will be:

$e_{8s} = (t_{e1} + t_{ts1}) + (t_{e2} + t_{ts2}) + (t_{e3} + t_{ts3}) + (t_{e4} + tt_4)$

where t_{ts}, t_{ts2} and t_{ts3} are the data transfer times for a sequential execution. There is an advantage in parallel processing of this algorithm

if $e_8 < e_{8s}$ i.e. if

$t_{t1} + \text{Max}\{(t_{e2} + t_{t2}), (t_{e3} + t_{t3})\} < t_{ts1} + (t_{e2} + t_{ts2}) + (t_{e3} + t_{t3})$

An implementation of the algorithm in a single computer or a two processor multiprocessor or a two station network will modify the values of the data transfer times and if the processors are dissimilar, the execution times.

This approach also allows data and control flow analysis. In the simple example fig. 5, we made the assumption that P2 and P3 use the same input vector for their computations. It could be that they require different inputs and then the subnet marked on fig. 5 would be:

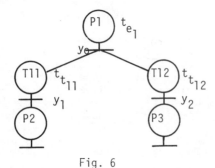

Fig. 6

In the case of fig. 5 the information that is input to P2 and P3 is the same y vector of the plant

$y = T_1(y_0)$

In the case of fig. 6 the information input to P2 is

$y_1 = T_{11}(y_0)$

and to P3 is

$y_2 = T_{12}(y_0)$

In general data transfer is a particular type of process with no memory i.e. the output does not depend on what has happened in the past.

$T = \{I, 0, A\}$

Flow analysis through a process place is more complicated because its outputs are function of their past process states. However the analysis of data flow through each process place is done as with standard software (Hecht, 1977) as process places do not contain real time features within them.

It is clear that a place representing a process in the system description can be analysed separately at another more detailed level of abstraction using graph (Petri Nets included) and software analysis tools.

EXAMPLE

Decomposition of a Multivariable Model Reference Adaptive Controller.

In this section we present the decomposition and analysis of a multivariable controller. Two decompositions will be analysed. First a decomposition based on the control functions is presented. The second decomposition is based on mathematical operations. The execution times of the purely sequential algorithm and the two decompositions are compared. The time that we will associate to the process places is expressed in a number of elementary operations (multiplications, additions).

The model reference adaptive controller with internal parallel model is presented in fig. 7.

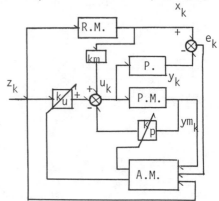

Fig.7 Model Reference Adaptive Controller.

Algorithm

The internal parallel model adaptive controller algorithm is (Nader, 1978):

1. Initialisations

2. Loop
   Reference Model
   $x_{k+1} = Am x_k + Bm z_k$

3. Parallel Model
   $ym_{k+1} = Ac ym_k + Bc u_k$

4. Plant outputs sampling
   y_{k+1}

5. New setpoint
   z_{k+1}

6. D.C. component supression
   $xo_{k+1} = x_{k+1} - xo$

   $yo_{k+1} = y_{k+1} - yo$

   $zo_{k+1} = z_{k+1} - zo$

   $ymo_{k+1} = ym_{k+1} - ymo$

   $eo_{k+1} = xo_{k+1} - yo_{k+1}$

7. Computation of vo_{k+1}
   $vo_{k+1} = D(eo_{k+1})$

8. Predictive Computation of vo_{k+2}
   $vo^o_{k+2} \leftarrow vo_{k+1}$ Approximation

   $vo_{k+2} = (I + DBp(N(k+1)))^{-1} vo^o_{k+2}$

   where

   $N(k+1) = (F+FP) yo^T_{k+1} G yo_{k+1} + (S+SP) zo^T_{k+1} Hzo_{k+1}$

9. Computation of $kp(k+1)$

   $kp(k+1) = Kp - \Delta kp(k+2)$

   where $\Delta kp(k+2)$ is calculated as:

   $\Delta kp^P(k+2) = FP vo_{k+2} (G ymo_{k+1})^T$

   $\Delta kp^I(k+2) = \Delta kp^I(k+1) + F vo_{k+2} (G ymo_{k+1})^T$

   $\Delta kp(k+2) = \Delta kp^I(k+2) + \Delta kp^P(k+2)$

10. Computation of $Ku(k+1)$

    $ku(k+1) = ku + \Delta ku(k+2)$

    where $\Delta ku(k+2)$ is calculated as:

    $\Delta ku^P(k+2) = SP vo_{k+2} (Hzo_{k+1})^T$

    $\Delta ku^I(k+2) = \Delta ku^I(k+1) + S vo_{k+2} (Hzo_{k+1})^T$

    $\Delta ku(k+2) = \Delta ku^I(k+2) + \Delta ku^P(k+2)$

11. Computation of uo_{k+1}

    $uo_{k+1} = -kp(k+1) ymo_{k+1} + ku(k+1) zo_{k+1} + km xo_{k+1}$

12. Close loop in 2.

    u, ym and $x \in R^n$ and $z \in R^L$

Evaluation of the Cost of the Sequential Algorithm

We neglect the data transfer times and consider only the process execution time costs. This is evaluated in terms of number of multiplications (M) additions without reference to the memory (i.e. register to register) (A) and additions with reference to the memory (AM).

The total execution time of the sequential algorithm is:

(N is the number of plant states and L of plant outputs).

(M) $2N^3 + 17N^2 + 6NL + 2L^2 + N + L$
(A) $2N^3 + 9N^2 + 4NL + 2L^2 - 14N - L - 2$
(AM) $6N^2 + 3NL + 9N + L$
    Sampling execution time
    Matrix inversion (N,N)

for detailed calculation of the execution time see appendix 2.

First Decomposition

Here the decomposition criterion followed is the conservation of control functional meaning for each process. The decomposition that respects this constraint with maximum parallelism is shown in fig. 8. Processes Pi correspond to the ith step of the algorithm. The total execution time of this decomposition is:

(M) $2N^3 + 16N^2 + NL + L^2 + N + L$
(A) $2N^3 + 8N^2 - 10N + NL + L^2 - 2$
(AM) $6N^2 + 8N + L$

Matrix inversion. (N,N).

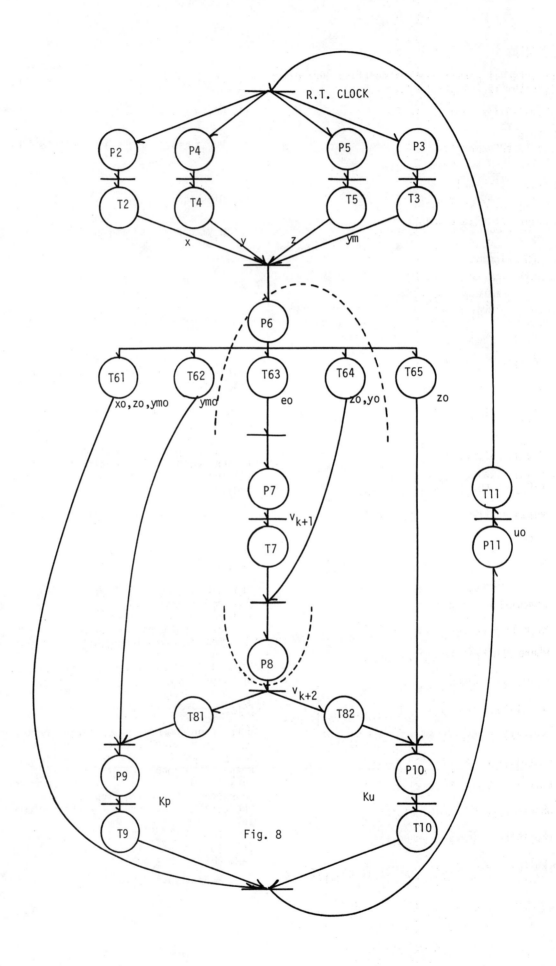

Fig. 8

The gain introduced by the first decomposition with parallel processing is:
(M) $N^2 + 5NL + L^2$
(A) $N^2 - 4N + 3NL + L^2 - L$
(AM) $N + 3NL$

Sampling execution time.

The total gain is small in relation to the total cost of the sequential algorithm especially if N is big ($N^3 \gg N^2$). In fact the maximum number of parallel processes is 4 (P2,P4,P5,P3). The major cost of the algorithm is the cost of process P8 that in this decomposition is purely sequential.

Second Decomposition

The decomposition criterion will be to consider matrix operations as elementary units in order to increase parallelism. One could consider decomposing the whole algorithm at the matrix operation level; this would produce an enormous number of processes and data transfers. The precedent decomposition has shown that the major execution time of the algorithm is due to process P8. We consider then the decomposition of P8 only, at the matrix operation level since the decomposition of the rest of the algorithm at this level will produce a marginal gain.

The subnet marked in fig. 8 can be decomposed into the subnet presented in fig. 9. The meaning of the different processes is presented in appendix 2. We have combined several processes in one place when they are purely sequential (e.g.P12 and P13).

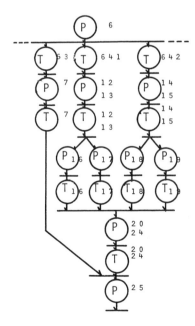

Fig. 9 Modified subnet for the second decomposition

The minimum execution time of the algorithm for the second decomposition is:
(M) $2N^3 + 12N^2 + NL + N$
(A) $2N^3 + 7N^2 - 9N + NL - 1$
(AM) $6N^2 + 8N + L$

Matrix inversion (N, N).

The gain of this second decomposition with respect to the sequential is:
(M) $5N^2 + 5NL + 2L^2 + L$
(A) $2N^2 - 5N + 3NL + 2L^2 - L - 1$
(AM) $3NL + N$

Sampling execution time.

The gain respect to the first decomposition is
(M) $4N^2 + L^2 + L$
(A) $N^2 + L^2 - N - 1$

In the two decompositions we have only taken into account the execution times. These are expressed in elementary operations to make the analysis computer implementation independent. The transfer data times have not been taken into account because they are dependant on the computer implementation (multiprocessor, network, etc.). In addition to the execution time of the processes there are the data transfer times.

$$t_3 + t_{63} + t_7 + t_{81} + t_9 + t_{11}$$

for the first decomposition and

$$t_3 + t_{642} + t_{14} + t_{15} + t_{19} + t_{20} + t_{21} + t_{22} + t_{23} + t_{24} + t_{81} + t_9 + t_{11}$$

for the second decomposition.

Note that the number of data transfers of the second decomposition increase rapidly although the data transfer time may decrease.

The problem of data transfer corruption can also be analysed and its impact on the control algorithm can be evaluated. Memory interference in a multiprocessor or traffic congestion in a network can also be analysed. Suppose that we want to implement this algorithm in a two processor architecture with no common memory configuration using the first decomposition. In this case (P4,P2) and (P5,P3) will be executed sequentially in each processor respectively. To minimize data transfer time P6,P7 and P8 should be implemented in the same processor that executes P5 and P3 (N>L). The process to processor assignment that minimizes data transfer is presented in Table 1.

Table 1. Assignment for the first decomposition.

Processor 1	Processor 2	Interprocessor Communication	Memory reference for processor 1
P5 P3 P6 P7 P8	P2 P4	2→1(X,Y)T4,T2 2 N real values	T5,T3 T63,T64 T7,T61, T62
P9	P10	1→2, (V,Z0) T82,T65 N+L real values	T81
P11		(2→1)(Ku)T10 NL real values	

CONCLUSION

Time Valued Petri Nets are a modification of standard Petri Nets with a particular software process interpretation. Time Valued Petri Nets are a simple tool for modelling, decomposing and analysing real time control algorithms for parallel processing. A great variety of aspects can be analysed with this tool: decomposition, data flow, concurrency, error propagation, execution time, interprocess communication, assignment of processes to processors, number of processors and interprocessor communication.

The application of Time Valued Petri Nets in modelling a life size control algorithm has shown the ease of application and the effectiveness of their use. It has been shown that they are a powerful tool.

APPENDIX 1

A Brief Introduction to Petri Nets.

A Petri Net is an abstract formal model for discrete systems. It is a mathematical structure that can be used for very different purposes for the definition of a system evolution. In particular they are very frequently used in the description of sequential systems and in the modelling of systems in which some events occur concurrently.

Definition

A Petri Net is a quadruple $N=(S,T,Z,Q)$ where S and T are finite not empty sets,
s.t. $S \cap T = \phi$ · (S-State or Place, T-Transition)
and Z and Q are binary relations s.t.

$Z \subseteq$ SXT forward incidence function
$Q \subseteq$ TXS backward incidence function

There exists an arc (flow relation) between the sate S_i and Transition T_j iff. $(S_i,T_j) \in Z$ and similarly there exist an arc between T_i and S_j iff. $(T_i,S_j) \in Q$.

Representation

The graphical representation of Petri Nets consist of two types of nodes. Places and Transitions are usually represented:

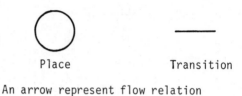

An arrow represent flow relation

→

and a token represents presence of state

•

Some interpretations of Petri Nets are:

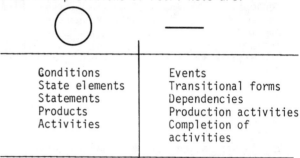

Conditions	Events
State elements	Transitional forms
Statements	Dependencies
Products	Production activities
Activities	Completion of activities

Tokens are contained in places. One or more tokens can be contained in one place. A place is marked if it contains at least one token. A marked net is a net with distributed tokens in its places. An example of a marked Petri Net is:

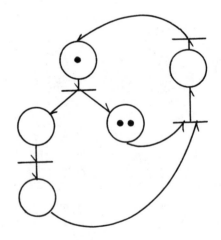

Petri Nets can also be represented in incidence matrices form.

A transition having several input and/or output places is called an 'AND' node.

General — Distribution — Junction

AND Nodes

A place having several input and/or output transitions is called an 'OR' node.

General — Selection — Attribution

Marking

The distribution of tokens in a marked Petri Net defines the state of the net and is called its marking. A mapping $M: S \to N \cup \{o\}$ is called a marking of a net N.
$M(s)$ is called the number of tokens of place $s \in S$.

Transition Activation

A transition $t \in T$ is called activated under a given marking if all input places of the transition t are marked.

Transition Firing

The firing of an activated transition occurs when the event associated with it occurs. This firing is defined as a transformation from one marking M to another M'.

The firing of an activated transition t removes one token from each input place and adds one token to each output place.

Examples of firing of transitions:

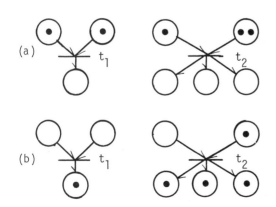

(a) activated transitions (b) firing.

Some Particular Petri Nets.

Pure petri nets. A Petri Net is called pure if the input and output places of a transition are disjoint sets.
   Example: Non pure Petri Net

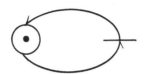

Live petri nets. A Petri Net is called alive for a given marking if for every attainable marking and for every transition there exist a firing sequence that allows the firing of this transition.
   Example: Non live Petri Net.

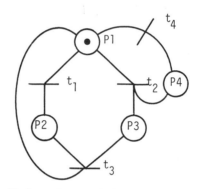

The firing of transitions t_2 t_4 t_1 t_3 presents a normal evolution. Any sequence starting with firing t_1 produces a deadlock.

State graphs. State graphs are Petri Nets without 'AND' nodes. Each transition has only one input and one output arc. Sequential systems are then easily assimilated to Petri Net description.
   Example:

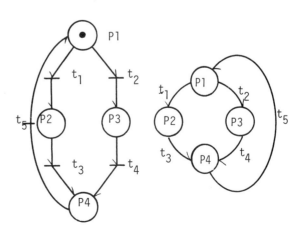

Petri Net   State Graph

Safe petri nets. A Petri Net is 'safe' for a given marking if for every accesible marking no place contains more than one token.
   Example: Non Safe Petri Net.

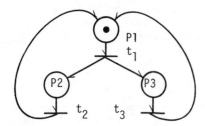

The firing of t_1 t_2 t_1 produces two tokens in P3.

The Advantages of Petri Nets.

We have seen that Petri Nets are very simple. They are used in a very large number of ways because they are a very powerful modelling tool. They are studied as formal automata or generators of formal languages. They are a major model for concurrent systems.

Petri Nets advantages are:
- powerful modelling tool
- formal
- ease of use
- analysable
- complexity and evaluation performance is possible
- allow the description of several levels of abstraction.

Transformations

Some Correct Transformations of Petri Nets are:

Some Applications Examples.

Example 1. Automaton.
Let C_1 and C_2 be two cars. C_1 moves between A and B and C_2 moves between C and D. r_i controls the movement to the right of C_i and l_i controls the movement to the left of C_i. Both cars start moving to the right if they are at A and C and if the control M is on.

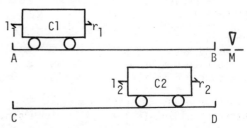

C_1 comes back to A after arriving at B. C_2 comes back only when C_2 is at D and C_1 has come back to A. The Petri Net model of this problem is:

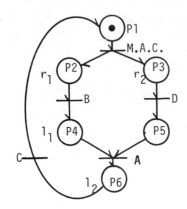

Example 2. Software modelling. Mutual exclusion.

Mutual exclusion arises whenever two or more processes require access to an unsharable resource. The resource may be unsharable, like a paper tape reader, because of its physical nature or, like a table memory, because simultaneous access by several processes might destroy the internal consistency of the data. The portion of a program executed by a process in accessing such a resource is generally referred to as a critical section and the mutual exclusion problem is that of ensuring that no two processes can be inside critical sections at the same time. That is if process 1 is executing its critical section process 2 must not begin its critical section until process 1 has left its own critical section. The solution of the mutual exclusion problem modelled with Petri Nets is:

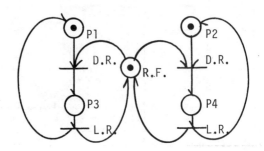

$P_1 P_3$ is process 1. P_3 is the critical section of process 1.
$P_2 P_4$ is process 2. P_4 is the critical section of process 2.
D.R. Demand Resource.
L.R. Liberate Resource.
R.F. Resource Free when P_5 is marked.

For further study see bibliographies in (Leutenbach, 1974) (Petri, 1975) (Peterson, 1977).

APPENDIX 2

Algorithm Execution Time Evaluation.

Cost of matrix operations: (M) multiplications, (A) additions and (AM) additions with memory reference.

Operation	Execution Time
Product of matrix (N,L) by a vector (L)	(M) NL (A) NL-N
Product of two (N,N) matrices	(M) N^3 (A) N^3-N^2
Addition of two (N,L) matrices	(AM) NL
Product of a vector (N) by a transposed vector (L)	(M) NL
Product of a transposed vector L by a vector L	(M) L (A) L-1

Execution times for each process in the multivariable controller.

Process	Execution Time
2	(M) $N^2 + NL$ (A) $N^2 + NL - 2N$ (AM) N
3	(M) $2N^2$ (A) $2N^2 - N$ (AM) N
4	Sampling time
5	0
6	(AM) $4N + L$
7	(M) N^2 (A) $N^2 - N$
8	(M) $2N^3 + 6N^2 + L^2 + N + L$ (A) $2N^3 + L^2 - N - 2$ (AM) $3N^2 + N$ Matrix inversion (N,N)
9	(M) $5N^2$ (A) $3N^2 - 3N$ (AM) $3N^2$
10	(M) $4NL + L^2$ (A) $2NL + L^2 - 2N - L$ (AM) $3NL$
11	(M) $2N^2 + NL$ (A) $2N^2 + NL - 3N$ (AM) $2N$

Execution times and process meaning for the second decomposition.

Process	Operation	Execution Time (M)	(A)	(AM)
P12	$Hzo \to a$	L^2	L^2-L	0
P13	$zo^T a \to b$	L	$L-1$	0
P14	$Gyo \to c$	N^2	N^2-N	0
P15	$yo^T c \to d$	N	$N-1$	0
P16	$Sb \to e$	N^2	0	0
P17	$SPb \to f$	N^2	0	0
P18	$Fd \to g$	N^2	0	0
P19	$FPd \to h$	N^2	0	0
P20	$e+f+g+h \to i$	0	0	$3N^2$
P21	$Bpi \to j$	N^3	N^3-N^2	0
P22	$Dj \to h$	N^3	N^3-N^2	0
P23	$h + I \to k$	0	0	N
P24	$k^{-1} \to 1$	matrix inversion (N,N)		
P25	$1V_{k+1}$	N^2	N^2-N	0

References

Allen, F. E. and J. Cocke (1976). A program data flow analysis procedure.
CACM Vol. 19, No. 3, pp 137-147

Baer, J. L. (1973). A survey of some theoretical aspects of multiprocessing.
Computing Surveys Vol. 5, No. 1, pp.31-80

Brantley, W.C., J.R., G. W. Leive and D.P. Siewiorek (1977).
Decomposition of data flow graphs on multiprocessors.
AFIPS NCC Vol. 46, pp 379-388.

Brinch Hansen P. (1977). Distributed Process a concurrent programming concept.
CACM Vol. 21, No. 11.

Fernandez, E.B, and B. Bussell (1973). Bounds on the number of processors and time for multiprocessors optimal schedules. IEEE Trans.Comput, Vol C-22, No. 8, pp.745-751.

Gaspart, P.and P. Vandenbussche (1978). An ergonomic and reliable system for distributed process control. IFAC 1978.

Hecht, M.S. (1977). Flow Analysis of Computer Programs. Elsevier North Holland. New York USA.

Hoare, C.A.R. (1978). Communicating sequential processes. CACM Vol. 21, No. 8, pp.666-677.

Horton, J.R.,I.R. James, A. Langsford and W. Teasdale (1977). A Real Time distributed processing system using GEC 4000 series computers. IFAC-IFIP workshop on real time programming. Eindhoven. pp.93-97.

Jenkins, D.W. (1978). Distribute or No? The choices between distributed and central computing control. Control Eng. Vol. 25, No. 6, pp. 61-64.

Kodres, R.U. (1978). Analysis of real time systems by data flowgraphs. IEEE Trans. Software Eng., Vol. SE-4, No. 3, pp. 169-178.

Kramer, J. and R.J. Cunningham (1978). Towards a notation for the functional design of distributed processing systems. IEEE International Conference on parallel processing. pp.69-76.

Kung, H.T. (1976). Synchronized and asynchronous parallel algorithms for multiprocessors. In J. F. Traub (Ed.), New Directions and Recent Results in Algorithms and Complexity. Academic Press, NY.

Lautenbach, K. (1974). Use of Petri Nets for proving correctness of concurrent process systems. In Information Processing 74. North Holland Publishing Co. pp. 187-191.

Mendelbaum, H.G. and F. Madaule (1975). Automata as structured tools for real time programming. IFAC-IFIP Workshop on real time Programming. Boston. pp. 59-65.

Nader, A. (1978). Modelisation et commande adaptative multivariable des caisses de tete de machine a papier. These Docteur-Ingenieur. Institut National Polytechnique de Grenoble.

Nader, A. (1979). The Design of Distributed Computer Control Systems. IFAC Workshop on distributed computer control systems. Tampa, Florida.

Paige, R.M. (1977). On partitioning program graphs. IEEE Trans Software Eng. Vol. SE-3, No. 6, pp. 386-393.

Parnas, D.L. (1971). Information distribution aspects of design methodology. Proc. of the IFIP Congress 1971. Ljubljana, Yugoslavia.

Parnas, D.L. (1972). On the Criteria to be used in decomposing systems into modules. CACM. Vol. 15, No. 12, pp. 1053-1058.

Peterson, J.L. (1977). Petri Nets. Computing Surveys. Vol. 9, No. 3, pp. 223-252.

Petri, C.A. (1975). Interpretations of Net theory. Internal Report. Gesellschaft fur mathematik und daten verarbeitung MBH Bonn Institut fur informationssystem forschung.

Raskin, L. (1978) Performance evaluation of multiple processor systems. Ph.D. Thesis. Carneige Mellon University. CMU-CS-78-141.

Schoeffler, J.D. (1978). Software architecture for distributed data acquisition and control systems. IFAC 1978.

Spang, H.A.III, (1976). Distributed computer systems for control. SOCOCO 1976. 1st IFAC-IFIP symposium on software for computer control. pp.47-65.

Stone, H.S. (1977). Multiprocessor Scheduling with the aid of network flow algorithms. IEEE Trans. Software Eng. Vol. SE-3, No. 1, pp. 85-93.

Stone, H.S. (1978). Critical load factors in two processor distributed systems. IEEE Trans Software Eng. Vol. SE-4, No. 3, pp. 254-258.

Syrbe, M. (1978). Basic principles of advanced process control system structures and a realization with distributed microcomputers. IFAC 1978.

Thomesse, J.P. (1977). A new set of software tools for designing and realizing distributed systems in process control. IFAC-IFIP Workshop on real time programming. Eindhoven.

ECONOMIC CONSIDERATIONS FOR REAL-TIME AIRCRAFT/AVIONIC DISTRIBUTED COMPUTER CONTROL SYSTEMS

B. A. Zempolich

Command, Control and Guidance, Research and Technology Group,
Naval Air Systems Command, Washington, DC, U.S.A.

Abstract. Using aircraft/avionic systems as examples, economic considerations for Distributed Computer Control Systems (DCCS) are discussed. Centralized, distributed, and federated processing architectures are used as the primary set of systems alternatives from which the economic factors are developed. Technical, schedule, and financial risks for the system architectures are presented. Standardization of computer hardware and software is examined from the economic viewpoint and other related risk factors. The economic impact of subsequent logistic support for standardized computer hardware and software versus non-standard products is identified. System considerations such as reliability, maintainability, availability, build-in-test, fault tolerance, and redundancy are examined from the standpoint of resources available to design and develop the DCCS, and also from the viewpoint of economic impact of failure of the DCCS to perform as expected. The economic impact of external factors such as the rate of technology advancement, technology independence, limited production runs, and the general lack of economic leverage upon the market are examined and related to the life-cycle support requirements of the DCCS. The economic considerations of a DCCS for an aircraft/avionic system application will be used to define similar considerations for an idealized automated process plant.

Keywords. Aircraft/avionic distributed computer control systems, system architectures, economic considerations; airborne computer system applications, architectures, control, hardware, interfaces, software, life-cycle support.

INTRODUCTION

The advent of commercially available solid-state electronics along with the discovery of the transistor started a revolution in aircraft/avionic equipment design. The aerospace industry, whether serving the interests of space, defense, or the commercial marketplace, quickly incorporated transistorized equipment into airborne applications. In particular, with the availability of the transistor, aircraft/avionic suites could, for the first time, incorporate digital computers into the system design without major technical or cost penalties. That the switch from analog to digital control took place with dramatic rapidity should not be surprising, as the ever-increasing needs of the aerospace operational environment constantly demand improvements in performance factors such as speed and throughput, and reduction in characteristics such as weight, volume, cooling, and power.

Since the early 1960's, the solid-state electronics industry has marched through a long series of quantum advancements in the technology -- e.g., first generation integrated circuits, then medium, large, and now very - large-scale integrated circuits. These advancements gave rise to the generic field of "microelectronics", from which both the microprocessor and microcomputer have been derived.

The inherent nature of microelectronic circuitry is that it lends itself to digital design techniques with ease. This attribute, coupled with the microprocessor, provides a powerful design capability to developers of aircraft/avionic systems. Today, powerful microcomputers can be embedded directly into each aircraft/avionic subsystem with little if any impact on weight and volume. With this new capability, top-down structured aircraft/avionic systems based on distributed processing architectures have become implementable and cost-effective.

This paper addresses the economic considerations which are a vital part of any decision-making solution to designing Distributed Computer Control Systems (DCCS) into future Avionic System Architectures. The aircraft as a DCCS is examined based on given aircraft mission and avionic system requirements. DCCS design options and alternatives ranging from physical implementations and alternative processing architectures, to standardization, commonality, reliability, maintainability and availability are analyzed from the economic viewpoint.

In addition to the options and alternatives available to the developing activity, there are many external factors which affect Avionic System Architectures over which the developer has little or no control. Among them are technology advancements, independence, limited production runs of microelectronic devices, and the general lack of economic leverage by the developers over the products of the solid state industry.

While precise economic parameters cannot be defined, general considerations and factors can be identified. From these an economic model can be created which identifies the important aspects of the economics of a Distributed Computer Control System for an aircraft. From this generalized model a comparison with an idealized automated process control plant is made.

AIRCRAFT MISSIONS AND AVIONICS REQUIREMENTS

The Aircraft As A Distributed Computer Control System

The modern aircraft is one of the best examples to demonstrate how a Distributed Computer Control System (DCCS) is used to perform system control functions in real-time. That this is so should not be surprising as aircraft have natural attributes which lend themselves to the incorporation of DCCS to perform on-board aircraft system functions. First of all, the requirement of an aircraft DCCS to operate in real-time is self-evident. That is, most jet aircraft travel at speeds near the speed of sound, while the Concorde and many military aircraft travel far in excess of the speed of sound. Secondly, to achieve the desired operational performance characteristics for any aircraft, weight and volume minimization of all on-board electric and electronic systems has to be a prime design consideration. Significant savings in weight and volume over today's conventional aircraft/avionic systems are projected with the use of distributed processing system architecture. Specifically, the use of microprocessors/microcomputers, coupled with standardized bus structures now permits the full realization of distributed processing in future aircraft design. Lastly, although there are many real-time, ground-based process control applications which require that the DCCS operate (geographically) in areas having hostile environmental conditions, in the majority of these applications the conditions tend to occur singularly rather than in combination with each other. On the other hand, the DCCS installed in an aircraft experiences a composite set of hostile environmental conditions at any one given point in time - temperature extremes with associated humidity and moisture, pressure variations, mechanical vibrations of both a steady-state and random nature, and electromagnetic interference over a broad frequency spectrum.

If the need for real-time operations, the availability of highly dense solid state digital technologies, and the capability to operate with physical dispersal of computer resources over a given geographical area are the keystones around which a DCCS is designed and developed for an aircraft, then it follows that the aircraft mission is the foundation upon which the building blocks are laid. For it is from the aircraft mission requirements that the overall blueprint for the total avionic systems (on-board electrical and electronic systems) is created.

Aircraft Missions

Primary avionic requirements are derived from the mission of the aircraft. For example, in the commercial airline industry there are two broad, basic missions for the aircraft - that of cargo carrying and that of passenger service. On the other hand, in the military environment, aircraft take on a number of highly individualistic and specific operational missions. Military requirements for aircraft also include the equivalent of cargo carrying and passenger service as in the private sector of the economy.

An aircraft mission is defined as the operational functions performed by the aircraft. The specific mission determines the type and capabilities of the functional subsystems (both electrical and electronic) on-board the aircraft. These subsystems may be "clumped" into several categories or groups. The vehicle and avionic core groups have common usage in both military and commercial aircraft, while the mission-sensors and weapons groups are unique to the military.

Examples of vehicle subsystems are: electrical, hydraulic, flight control, and propulsion. Core subsystems are those avionic equipments which perform functions such as: communications, navigation (including radar), and general data processing. The military requirements for mission-sensors are for those equipments which perform functions such as the following: radar, infrared detection, low light level television (LLLTV), and operator displays. Examples of weapon subsystems are guns, bombs, missiles, rockets, stores management units, and pylons/launchers.

Avionic Systems Requirements

Once the mission of the aircraft is established, the avionic systems requirements are defined using a structured design approach. Specifically, the avionic systems requirements are divided into subsystem structures (applications) using a top-down decomposition of the mission requirements. That is, the mission requirements are decomposed into application requirements for the subsystem categories described earlier - vehicle, core, mission-sensors, and weapons.

Using the structured design approach these various subsystems are iteratively broken down (decomposed) into modules which can be implemented as either hardware, firmware, or software. Next, the hardware components are selected and a hardware configuration determined. In turn, the firmware and software are mapped onto the hardware configuration.

Avionic System Architecture

In order to have the proper interrelationships between the aircraft/avionic subsystems and their constituent parts, the structured design approach must have an initial overall system architecture upon which the decomposition process is performed. By definition, the Avionic System Architecture includes not only the processing partitioning between the various subsystem categories, but also the associated system integration and hardware commonality considerations. Table 1, System Architecture Considerations, provides a comprehensive listing of functions which are all interrelated in some fashion and must be examined under the aegis of the Avionic System Architecture definition.

TABLE 1 System Architecture Considerations

- Processing Partitioning Between Subsystems
- Interconnect Schemes
- Interface, Control and Protocol Standards
- Fault-Tolerance, Reconfigurability and Built-In-Test
- Reliability, Maintainability and Availability
- Man-Machine Relationships
- Automated Operator Decision Aids
- Systems Integration
- Hardware and Software Standardization
- Commonality of Equipments
- Technology Insertion and Independence

ECONOMIC CONSIDERATIONS

DCCS System Design Options and Alternatives

As with most engineering efforts, the design of an aircraft DCCS allows the developer to exercise a number of options, all of which have interactive technical, schedule, and economic (cost) risks. DCCS design options and alternatives generally fall into two categories - those factors over which the designer has direct control and those factors over which there is little or no control by the developing activity. DCCS considerations over which the developer has control include: physical implementations; alternative processing architectures; standardization and commonality; and reliability, maintainability, and availability.

Physical Implementations

As stated previously, once the primary mission for an aircraft is established, the on-board Avionic System Architecture can be decomposed into the various functional requirements. In a similar fashion, subsystems can be partitioned into various physical implementations. There are three basic equipment physical implementation alternatives: The "black-box" approach; the form, fit, function (3F) approach; and the integrated technologies concept.

With the black-box approach, all equipment procurements over the life-cycle of the aircraft are bought to a set of specifications which detail not only the function and form, but also the internal configuration - electronic, electromechanical, and packaging. Once the desired performance of the unit is established, subsequent procurements usually have minimum technical and schedule risk. Quantity of units to be bought per unit of time is the dominant economic factor with procurement of black-box implementations of avionic equipments. Multiple suppliers can also be considered a major force in price determination as the competitive atmosphere tends to keep the cost of the equipment down. Additionally, long-term logistic considerations (which have a great impact on the life-cycle costs of the aircraft) can be established after the equipment reaches development maturity.

A second physical implementation alternative for avionic equipments is that of form, fit, and function (3F). With the 3F approach, procurements of equipments are made to a set of specifications which determine the physical dimensions as well as the electronic and electromechanical interfaces. The technologies of the assemblies within the unit are allowed to vary or "float". The economic value of the 3F approach rests mainly with the options open to the supplier in having to meet only the 3F specifications. In essence, the supplier is free to make maximum use of his resources and manufacturing facilities. It is normal to expect that there is the potential for cost savings through the use of the 3F approach in that it opens the door for more suppliers to bid. However, there is an economic shortcoming of the 3F approach in that it does not readily lend itself to long-term logistic gains and planning.

In both the black-box and 3F approach each avionic unit performs a fixed, specific function. At the other end of the spectrum, the Avionic System Architecture can be partitioned along the lines of integrated technologies in which functions are performed by generic task areas such as data processing, communications, navigation, or controls and displays. In this instance, advanced technologies are used in an integrated fashion such that any one given part of the subsystem is capable of performing different functions at different times. Specifically, with this implementation, the elements are all electrically reconfigurable.

While this concept has considerable technical, performance, and economic merit, it has yet to be fully exploited in avionic applications, and thus the risks are not yet well established in terms of specific parameters.

Regardless of the alternative selected, the physical implementation of the aircraft/avionic equipment(s) is a fundamental design decision which has major technical and management impact during the development phase and during the operational life of the aircraft. For this decision dictates basic logistic support approaches for the system such as depot repair, module "throw-away" concepts, or factory repair and maintenance.

If the decision regarding which physical implementation alternatives should be selected could be made solely on the considerations addressed in this section, the choice is reduced solely to a comparison of risks. Unfortunately, the choice is also dependent to a large degree on the aircraft installation itself. Specifically, is the installation of the DCCS to be made in an existing operational aircraft as opposed to an installation in a new airframe? With a new airframe, the weight, volume, and location of the equipment is normally determined concurrently with the development of the aircraft; thus there is a degree of design latitude allowed in the integration of the aircraft/avionic subsystem. On the other hand, with an existing airframe, there are a number of significant restrictions on the installation of a newly designed DCCS because of existing conditions.

The importance of installation options cannot be overstated. Restrictions that may have to be faced with existing aircraft may well prevent an optimal combination of airframe and on-board aircraft/avionic subsystems from a logistic viewpoint. Needless to say, logistics considerations are in reality economic considerations, and if experience to date is any measure, the costs for lifetime logistical support far exceeds the non-recurring development costs.

Alternative Processing Architectures

The modern avionic DCCS will be required to handle a wide variety of tasks ranging from complex, high speed signal processing to simple input/output formatting and control. Additionally, fault tolerance concepts demand that many of the processing elements within the DCCS be capable of reprogramming during the operational mission. The overall processing architecture must therefore support the synchronization, control, configuration, reconfiguration, and fault detection of all processors in the DCCS. Furthermore, to minimize architectural problems, both the hardware and the software must be functionally partitioned in such a manner that the interface complexity is manageable, and the design and implementation of each unit processor is maintained in as independent a manner as is possible.

There exists a variety of processing architectures which can be utilized to design an aircraft/avionic DCCS with the performance capabilities just identified. It should be noted, however, that each alternative has attached to its use a unique set of technical schedule, and financial risk factors. Figure 1, Processing Architecture Alternative Comparison, lists a number of available processing architecture options and identifies the associated risk factors. Risks are stated in low, medium and high terms because there is presently a lack of a statistical data base from which precise numerical values can be derived.

Unfortunately, the procedure for selecting a specific processing architecture is not solely a matter of looking at the risk factors inherent in the individual architectures and determining what is an acceptable composite level of overall risk to developer. For example, the Avionic System Architecture Considerations identified in Table 1 also weigh heavily upon the decision concerning which processing architecture is "best" for a specific application. The necessity for having to take into consideration both the processing architecture alternatives as well as other Avionic System Architecture factors provides the developing activity with a myriad number of possible combinations from which to choose during the design of the DCCS. The technical management task required to separate these combinations into a set of hierarchically structured options based upon a well understood set of selection criteria is complex unto itself.

Because of the large number of interrelated factors which affects the selection of a processing configuration for a specific Avionic System Architecture and the lack of historical cost data base, one can only address in general terms the economic considerations of the various processing alternatives. Even though these economic considerations are addressed in general terms, they should not be interpreted as being either superficial, lacking in importance, nor restricted to only one architectural choice. For even as incomplete as is the cost data at this point in time, trends can be drawn from experiences with the individual requirements of current aircraft/avionic systems. Examples of considerations which have significant impact upon the life-cycle cost of a DCCS and require detail management attention by the developing activity during the project planning phase are: degree of system integration; degree of partitioning of the system; software, firmware, and hardware trade-offs; and software cost/complexity.

Degree of System Integration. This issue addresses the degree of total system integration of the Avionic System Architecture. For example, should the categories or groups of subsystems identified earlier be placed on a single high-speed data bus or should

ARCHITECTURE ALTERNATIVES	RISK		
	TECHNICAL	SCHEDULE	FINANCIAL
(1) DEDICATED SUBSYSTEM PROCESSORS	LOW/MEDIUM	LOW/MEDIUM	LOW/MEDIUM
(2) REDUNDANT DEDICATED SUBSYSTEM PROCESSORS WITH LOCAL BUSES	MEDIUM (WEIGHT)	LOW	LOW/MEDIUM
(3) REDUNDANT DEDICATED SUBSYSTEM PROCESSORS	MEDIUM/HIGH (WEIGHT)	LOW	LOW
(4) REGIONAL GROUPS OF SUBSYSTEMS	MEDIUM	LOW/MEDIUM	LOW/MEDIUM
(5) CENTRAL PROCESSORS	MINIMUM	MEDIUM (INTERFACING)	HIGH (INTERFACING & SUPPORT)
(6) MULTIPROCESSORS	HIGH (SOFTWARE & BUS PROBLEMS)	MEDIUM/HIGH	MEDIUM/HIGH

Fig. 1. Architecture alternative comparison

each group have its own dedicated data bus to perform functions particular to the individual grouping of subsystems. A specific example of the dedicated data bus would be to keep all vehicle-related subsystems segregated for safety-of-flight reasons. It can be anticipated that if there is one high-speed data bus throughout the aircraft, then the complexity of controlling the data bus and performing real-time executive and interrupt functions would be increased dramatically. In turn, software-related costs (design, test, and documentation) would increase significantly, if not proportionately with the degree of integration. This conclusion is based on the fact that cost experience with operationally deployed aircraft systems to date has shown that the real-time executive and I/O routines are much higher than application programs and test and diagnostic routines on a cost (in dollars) per instruction.

Degree of Partitioning of the System. As stated earlier, future aircraft DCCS's must be designed using a structured process of decomposition into software, firmware, and hardware processing modules. For future aircraft the degree of distribution (partitioning) of computing, control, and conversion functions, will be dependent upon the availability of inexpensive and physically diminutive hardware elements - namely, microprocessors and microcomputers. It should be noted however, that while the use of a central computer complex to provide functional digital control of an aircraft has deficiencies due to the multiplicity of tasks which must be performed in one machine, the DCCS has yet to face the same problem while performing similar tasks with as many as up to 150 to 200 (micros) machines.

Software, Hardware, and Firmware Trade-Offs. Analog computers have traditionally performed fixed functions and inherently do not have the capability of change without an attendant hardware impact. The programmable digital computer added the dimension of permitting inservice functional change without impacting the associated hardware, except where additional memory was required. With the introduction of firmware, the "best of two worlds" is available. The options for committal of functions to firmware implementation as opposed to software is unbounded in number. Key to any decision-making process as to whether or not to put a function into firmware is when should one freeze the software program design, and how often, if ever, is the program going to be required to be changed throughout the operational lifetime of the system. Any misjudgment on the proper timing for freezing the program into firmware and miscalculation on the number of times that the firmware will require change, will result in major increases in development and support costs.

Software Cost/Complexity. In the centralized processing architecture, the cost and complexity of applications/control and Input/Output programming rises exponentially as the throughput and memory of the centralized computer approaches its maximum (see Fig. 2). On the other hand, with the distributed processing architecture, the cost/complexity at near zero percent (0%) distribution is the same as one hundred percent (100%) utilization of a centralized computer system. As the degree of distribution (i.e., partitioning) is increased, each application software module becomes more independent and has less effect on the execution of the total onboard system processing (program). The I/O program, however, becomes more complex since more processing elements (micros) must be interfaced via the data bus structure. The data availability and I/O control becomes the dominant factor, ultimately following the I/O program curve of the centralized computer system in rising cost/complexity (see Fig. 3). The sum of the software trends indicates that there is probably a point at which partitioning may be optimum. As is self-evident from Fig. 3, that at either end of the distribution spectrum, the worst of both worlds may exist.

Standardization and Commonality

It is this author's opinion that no other area of the data processing field is more complex in scope and controversial in nature than the area of standardization. Many professionals in the field of data processing do not agree that standardization has both technical and cost merit. This lack of consensus on the worth of standardization is due to the naturally opposing views of computer system users and the developers of computer systems. For the user views standardization as a means of management control of development risks and system life-cycle cost control; while the developer and designer, on the other hand, view standardization requirements as an unnecessary restriction on technical creativity. Many developers also counter the user's position that proliferation of computer equipment and software is a major life-cycle cost burden with the claim that given design freedom during the development phase of a new system, they would introduce new technologies which would be both cost-effective as well as having increased performance capability over existing operational systems. Unfortunately, there is a tendency amongst proponents of this development philosophy not to mention that new designs also give rise to normal self-vested interests, such as increased profits and keeping the in-house design teams current with involvement in emerging technologies and techniques. These two diametrically opposed positions will never change, in this author's opinion, as the developer deals with the technical and financial aspects of the specific system he is developing; while the user is concerned with standardization as applied to multiple system applications. Also, there is another dimension to the standardization issue which often is not considered in any discussion of computer systems standards. Specifically at what point or level does one standardize? For example, one could standardize at the

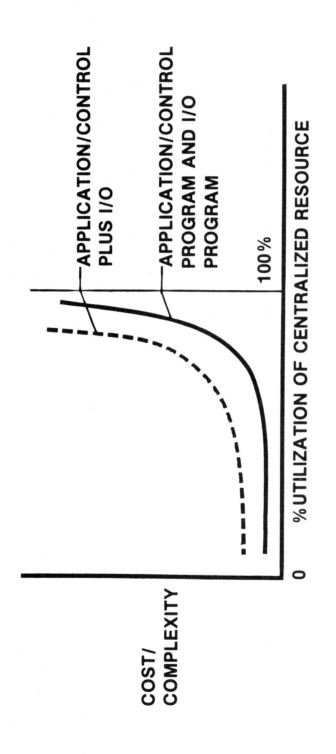

FIGURE 2: CENTRALIZED PROCESSING ARCHITECTURE SOFTWARE COST/COMPLEXITY

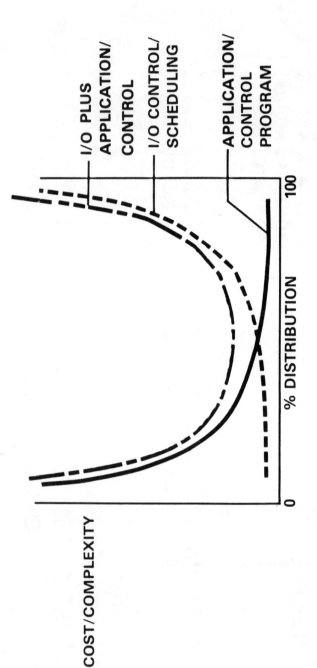

- THE SUM OF THE TWO SOFTWARE TRENDS INDICATES A POINT OF DISTRIBUTION WHICH MAY BE OPTIMUM. FURTHER, AT EITHER END OF THE DISTRIBUTION SPECTRUM THE WORST OF BOTH WORLDS MAY EXIST!

FIGURE 3: DISTRIBUTED SYSTEM TRADEOFFS

Instruction Set Architecture (ISA) level while allowing the designer to incorporate the latest technologies and change the physical and electrical characteristics -- e.g., overall dimensions, the internal mechanical structure of the machine, and cooling and primary power requirements.

Table 2 "Standardization Options" lists a number of possible standards which the user and/or the developer of aircraft avionic equipment could adopt. Several or many of these options could be combined to form an all-encompassing single standard depending on the financial resources available, maintainability/support approaches, and the end operational use of the system(s). However, the more these standardization options are melded into one single standard, the greater will be the negative reaction of the developer as stated earlier.

TABLE 2 Standardization Options

Languages
- Preprocessor (POL)
- Compiler (HOL)
- Assembler (MOL)

Instruction Set Architecture (ISA)
- Single Instruction Set
- Modular Instruction Set
- Extensible Instruction Set

System-Level Interconnection Schemes
- Bus
- Loop
- Network
- Bus Interface Unit

System-Level Protocol
- User Module to Operating System
- Operating System to Hardware

Physical Interface
- Pin Compatible
- Plug Compatible

Physical Implementation
- Black Box
- Form, Fit, Function
- Standard Module
- Micro-chip Set

Of all the Standardization Options listed in Table 2, adoption of an Instruction Set Architecture (ISA) as a standard offers the greatest economic return on investment. This is assuming that the ISA selected as a standard has an established user and support software base.

If one were to address the standardization issue solely on the basis of generalized hardware and language (HOL) alternatives, then a matrix of comparative risks can be defined. Figure 4, Hardware Standards, shows the technical, schedule, and financial risks for various hardware alternatives. It should be noted that high and medium/high risk factors have been assigned to the Strict Processor and Microprocessor Standards because of: (1) the lack of experience with building DCCS's for aircraft/avionic systems applications; and (2) it is not clear at this point in time that a single, cost-effective microprocessor can be established as a standard for all applications throughout an Avionic System Architecture.

The key issue relative to establishing a micro as a standard piece of hardware is at what point does one not enforce standardization. For example, is every application which calls for a micro whose word length is less than 16 bits subject to the standard? Or, is there a minimum memory size below which the micro would be excluded from standardization considerations? These decisions, while seemingly inconsequential, do have a significant impact on the design of the system and development costs.

Many individuals have postulated that microprocessors will decrease the cost of computer hardware to the point at which it is an insignificant factor on future developments of DCCS's. This claim has yet to be proven. Unfortunately, the rising costs of both applications and support software have lent credibility to the position that the cost for micros are no longer of relative importance in system life-cycle cost considerations.

Regardless of the availability of comparatively low-cost microprocessors and microcomputers, the high cost of software development and maintenance has given considerable support to the utilization of HOL's and, in particular, a single HOL wherever possible. Figure 5 shows the comparative risk factors for several language alternatives. Figure 5 indicates that assembly level coding is definitely more costly than that of using HOL(s). There are two major reasons for this cost differential: (1) there is the need for the programmer to know the particular Instruction Set Architecture of the target machine(s); and (2) in most cases assembly level code is used mainly for very difficult program tasks such as: input/output, operating systems, and executive control of real-time systems. In each of these instances the programmer must work with "tight" coding requirements.

Within the context of this paper, commonality is defined as the utilization of equipment(s), or parts thereof, in multiple operational applications. For example, many aircraft cockpit controls and displays could be common within a single "family" of aircraft types. Each aircraft, however, would have a specific set of cockpit controls and displays tailored to its own particular operational need. Across all aircraft within the family, the controls and displays would perform common functions. The equipment itself need not be standard items to be considered within the context of commonality as the term is used herein.

The potential for cost-savings does not exist with common equipment as it does with standard equipment because of the specific tailoring or uniqueness to each application. On the other hand, when the developer applies

HARDWARE ALTERNATIVES	RISK		
	TECHNICAL	SCHEDULE	FINANCIAL
STRICT PROCESSOR AND MICROPROCESSOR STANDARDS	HIGH	MEDIUM	MEDIUM/HIGH
PROCESSOR STANDARDS ONLY	LOW/MEDIUM	LOW	MEDIUM/HIGH
NO STANDARDS	MINIMUM	HIGH	HIGH

Fig. 4. Hardware standards

LANGUAGE ALTERNATIVES	RISK		
	TECHNICAL	SCHEDULE	FINANCIAL
SINGLE HIGH LEVEL LANGUAGE	LOW/MEDIUM	LOW	LOW
MANY HIGH LEVEL LANGUAGES	MEDIUM	MEDIUM	MEDIUM/HIGH
HIGH LEVEL LANGUAGE AND LOW LEVEL CODING (FALLBACK)	LOW	MEDIUM/HIGH	HIGH

Fig. 5. Language standards

commonality concepts effectively, there is a great potential for significant cost-avoidance. For example, specific display components, bulk memories, algorithms, etc., are applied across all applications. In doing so, the developer avoids those costs associated with developing totally unique equipment designs for each installation.

Reliability, Maintainability, and Availability (RMA)

In simplistic terms, aircraft/avionic systems are designed to meet pre-established levels of reliability so as to be available for operational use for given time-periods prior to a failure occurring which would require maintenance action to be taken. Where the reliability levels are not achieved, the equipment is not available and additional maintenance actions have to be taken. This cause and effect situation is a major contribution to operational support costs for both military and commercial aircraft. In the author's opinion, it is highly unlikely that with the current degree of technical sophistication of aircraft/avionic equipment these costs will decrease in the near future. Furthermore, unless new Avionic System Architectures are developed and designed as described earlier, the current RMA problems will remain.

It should be emphasized that using a DCCS as the basis for a future Avionic System Architecture will not of itself negate the current RMA problems; however, if the system is designed in a structured fashion, it can include many features which would assist in reducing RMA shortcomings as exhibited by current operational systems. Key features which, if properly incorporated into the design of the DCCS, will have a major impact in improvement of RMA factors and a corresponding reduction in life-cycle operational costs are: fault-tolerance, redundancy, and reconfigurability.

The capability to incorporate fault-tolerance, redundancy, and reconfigurability techniques and concepts into a DCCS is based primarily on the availability of relatively inexpensive micros. Given that these micros will be available, the major question remaining is at what level does the developer insert these concepts into the design of the DCCS. For these concepts can be applied either on a system-wide basis, or at any of the subsystem or functional grouping levels. Furthermore, with the coming of age of the reconfigurable memory, one can now have increased availability at the component level.

The coupling of fault-tolerance, redundancy, and reconfigurability with automated fault-detection and isolation also offers management a vehicle for minimizing RMA life-cycle costs of future DCCS's. Unfortunately, the expected theoretical improvements in the RMA values have yet to be fully proven out in actual practice over a substantial period of operational time. While there is no reason to believe that the potential gains cannot be achieved, there is an area of concern (mentioned earlier) that should be addressed during the development of the Avionic System Architecture -- namely that of the actual amount of distribution of computing resources throughout the system and its impact upon the software.

The complexity of the software associated with a DCCS is going to be a major challenge by itself. There are many problems yet to be faced with an aircraft/avionic DCCS which may contain over 150 micros. Additionally, there could be hidden costs because of unforeseen needs for performing extensive test and evaluation of such a system. Hopefully, sufficient software verification and validation techniques will be available to insure that the developer can adequately separate proving the quality of the software from the quality of the DCCS to function adequately as an integrated network of computer resources.

EXTERNAL FACTORS IMPACTING DCCS DEVELOPMENTS

External Factors

It took nearly three decades for the commercial computer industry to develop what has been generally termed as 3-1/2 or 4 generations of computer systems. For the purpose of this paper, the actual number depends on one's definition of what constitutes a "generation". A generation is defined herein as a significant improvement in speed and major increase in memory capacity. By comparison, within this decade the microprocessor has gone through numerous generations, again using speed and memory density as the basis for comparison. It is fairly self-evident that solid state technology advancements have been the driving force behind these quantum jumps in performance capability rather than carefully planned development efforts by the classical mainframe computer manufacturer. Regardless of whether the commercial computer manufacturers or the solid-state electronics firms are controlling the marketplace, collectively they have dramatically changed the design and development of computer systems within the decade of the seventies. One has only to realize that the concept of a "computer on a chip" is no longer a figment of some person's imagination to fully appreciate the dramatic changes which have occurred in the industry within the past few years. All of which leads to the question that developers of a DCCS must ask themselves before starting out on a new design; namely, just what degree of control do they have over their final designs. Unfortunately, the dynamics of the microelectronics industry as mirrored by the microprocessor/microcomputer marketplace defy the providing of reasonably precise answers to the question. At best, one can only hope that the impact upon the DCCS development efforts and related life-cycle considerations of the Avionic System Architecture are minimized through the recognition of these external factors during the planning phase of the project. The following external factors are identified as having major impact upon the DCCS design and development and thus should be addressed during the planning phase of the project: technology advancement, technology independence, limited production runs as a function of time, and lack of leverage upon the market, and the vertical structure of certain corporations.

Technology Advancement

It is almost inconceivable that the technological inventiveness of the solid-state electronics industry is such that new products become obsolete almost immediately after introduction into the market place. Breakthroughs in such areas as materials, manufacturing processes, computer aided design, architectures, and packaging are made almost daily. Furthermore, it is highly unlikely that in the near future there will be any slow-down in new performance capabilities being introduced in the microprocessor/microcomputer marketplace. If anything, there will be a continued explosion of new applications as the prices of these machines (micros) decrease as a function of time.

All other design factors being equal, advancements in the solid state electronics field are not necessarily detrimental to the aircraft DCCS developer. Desired system-level capabilities such as redundancy, reconfigurability, and fault-tolerance can now be built into the system economically and contribute to achieving the desired performance goals set for system maintainability, reliability, and availability. On the other hand, these capabilities cannot be logistically supported over the life-cycle of the system DCCS without taking into account the other external factors which impact DCCS developments.

Technology Independence

In a similar manner to the now famous commercial computer industry expression of "plug-to-plug" compatibility, the phrase "technology independence" has been introduced into the lexicon. In a manner of speaking, it can be considered a technology level equivalent to the form, fit, and function physical implementation approach addressed earlier. The concept is very simple, that is, by being independent of technology uniqueness, one can insert new technologies at given time intervals during the life-cycle of the aircraft/avionic DCCS. The economic return on investment for incorporating this capability into the system design is significant. On the other hand, it does demand that there be some level of mechanical packaging standards in order to introduce the new devices into the existing equipments with minimum impact upon logistic considerations. Assuming that a standard mechanical packaging concept can be established for both the current and replacement technologies, then there will be a logistics cost avoidance in that the higher level electronic assemblies do not change with the insertion of the new microelectronic technology.

On the other hand, with software, technology independence takes on a number of meanings, all of which depend on the point of view of the developer. For example, the applications and support software for a given programmable digital computer could be run, with no changes, on a newer technology machine providing the Instruction Set Architecture and other software program dependent characteristics are taken into consideration during the initial design phase. A second concept would be to keep the High Order Language (HOL) interface independent of the target machines. A third approach is that of using a pre-processor in the software development chain. Specifically, with this approach one establishes the near-equivalent of a hardware plug-to-plug compatibility by using a pre-processor as a software program translator. In this instance, the firmware is used to provide the software compatibility link.

Regardless of the type of hardware technology used, the concept of software transportability implies unto itself, technology independence. However, unlike hardware technology independence, software transportability of its very nature explicitly implies reusability of software as opposed to the basic idea behind plug-to-plug compatibility; namely, that of technology insertion through technology invisibility.

It can be stated that software transferability offers the developer a basis for cost savings. On the other hand, since the applications/programs will no doubt be different to a certain degree from functional task to functional task, new compilations will have to be performed in order to insert different application dependent parameters and data. Thus it is perhaps more correct to state that as a minimum, using software transportability concepts in an aircraft/avionic system design there will be a cost avoidance in that both the operational and support programs do not have to be created all over from the beginning.

Limited Production Runs

There is no better method to insure price stability than that of having the advantages that accrue from large scale procurements over a given period of time. In essence, this is the economy of scale factor of classical economic theory. Unfortunately, it is a fact of life that at best there will be limited quantities of aircraft/avionic Digital Computer Control Systems procured by any one development/procurement activity. Even if an aircraft manufacturing firm has incorporated DCCS's into several different aircraft models, the quantities of aircraft coming off the production line are miniscule compared with the quantities of microelectronic chips currently being procured by the automotive and the toy industries on a per year basis.

It would appear that there are two management alternatives which would overcome the inherent economic shortcomings of limited production runs. The first approach, would be to add onto existing production runs which are expected to produce microelectronic chips over an extended period of time. In this instance, individual procurement of chips for the DCCS would be made part of a standard product line which the solid-state electronics firms expect to market to multiple users into the foreseeable future.

In the second case, the aircraft/avionics systems manufacturer would "front-end" the development costs associated with the design of a given microelectronic chip and only use the solid-state electronics firms as a production facility. Thus the system developer orders parts to his specifications and is not dependent upon the microelectronic circuit manufacturers for any initial non-recurring investment in chip design and development costs.

It is essential that an acceptable alternative be established prior to production in order to maintain the availability of chips throughout the lifetime of the DCCS or until the chips are replaced by a newer technology during the operational stage of the system deployment. It is imperative to note that the lack, or shortage, of logistic spare parts destroys any logistics planning performed during the R&D stage of the DCCS, and further compounds the subsequent operational problems which range from system availability to maintainability.

Lack of Economic Leverage

Since World War II, the aerospace industry (both military and civilian) has introduced many advancements in the electronic state-of-the-art into the operational usage environment. In general, the industry introduced new technologies because they have had both the need as well as the economic leverage to do so. Over the last decade, this preemptive position has been eroded until at the present time the aircraft/avionic developers have very little impact upon the technical directions of the solid-state technology industry (based on a percentage of sales). Neglecting such global considerations as macroeconomics, the changing role of the multinational firms, and the emergence of a truly international capability to manufacture solid-state electronic devices, no single factor has had such a major negative impact upon the economic leverage of the aircraft/avionic firms over the solid state electronics marketplace as that of the coming of age of microelectronic circuitry. The aerospace firms first introduced the integrated circuits into aircraft/avionic applications in the early 1960's. Since then, application of integrated circuits permitted developers to design aircraft/avionic subsystems having performance parameters never before achievable within the weight and volume constraints of the operational environment

Since the mid-1960's, the combined sales of aircraft/avionic systems to both the private and public sectors has declined. While decreasing sales volume of aircraft per unit of time has had a profound effect upon industry leverage, it has really been the quantum jumps in densities of the chips (transistors per unit of area) which has become the dominant factor in changing who has the leverage over the industry. That this is so should be somewhat self-evident in that the higher density chip made obsolete the first generation "integrated circuit". It was, for all practical purposes, a single (physical) low cost replacement for hundreds of individually packaged integrated circuits. In reducing by orders of magnitude the number of chips to be procured, all vestiges of economic power over the marketplace by the aircraft/avionic system developers disappeared.

In retrospect, it is somewhat ironic that in the early 1960's it was the aircraft/avionic industry which was the only group of users that "carried" the infant microelectronic industry during those days of integrated circuit venture enterprise. By contrast, today a common 3 to 5 chip microcomputer design serves applications in the aerospace, automated factories, medicine, as well as the home entertainment markets equally well. On the other hand, projecting for the future, there is the possibility that there may be yet another "role reversal" concerning leverage in the market. Specifically, the use of Very Large Scale Integrated Circuits (VLSIC) in aerospace applications may very well prove to be the key factor in having the microelectronic circuit manufacturers re-tooling to meet once again the initially unique needs of the aerospace industry. Whether this situation will come to pass has yet to be determined. Until that time, however, aircraft/avionic systems developers will have to fit their needs into the standard product lines of the solid-state electronics industry if they do not wish to incur large non-recurring costs for customized chips.

Vertically Structured Corporations

Throughout the private sector there are many instances where a corporation is vertically structured - that is, where the organization is made up of companies and/or divisions which supply the raw materials, engineering (including R&D), manufacturing, and sales and distribution functions. In essence, the corporation does not go outside of itself for any major aspect of its operations and for all practical purposes is its own supplier of goods and services. The "verticality" of the organizational structure is derived from the nature of the manufacturing process whereby a unit of the corporation builds upon the output of another part of the organization. The management and cost advantages of this situation whereby availability of materials, scheduling, and commitment to corporate goals are all self-contained and controlled needs no further amplification.

With the advent of the transistor, many firms added a solid-state division (as a separate profit and loss center) to the corporate organization. Except in certain instances, the majority of these solid-state plants manufactured parts for the general commercial marketplace with no objective of serving internal corporate needs for devices such as transistors. In the author's opinion, the subsequent introduction of the microelectronic chip initiated the push for many aircraft/ avionic equipment manufacturers to also take corporate action to change to a vertical organizational structure. For the microelectronic chip took away many design prerogatives from the developers, and effectively made the solid-state electronics manufacturing firm a design competitor, albeit at the very low end of the design process. However, as the techniques for manufacturing microelectronic chips matured, and the industry introduced medium and large scale integrated circuits, the impact upon the classical design freedom of the aircraft/avionic equipment firms became fairly significant as the chips began to contain more and more of the individual circuits previously developed as physically separate designs.

To counter the growing impact of the external factors addressed earlier and the inroads that advanced microelectronic circuitry was making upon their traditional development efforts and organizational makeup, many aerospace firms changed their corporate structures to a vertically-oriented one. What many of these corporations did within the past decade was to create an in-house solid-state organization with the prime customer being the corporation itself. The capabilities of these in-house facilities are, as could be expected, as sophisticated and advanced as many of those in California's "silicon valley".

It is premature to state that the vertically-oriented aerospace firm will provide a management approach to overcoming the negative aspects of external factors such as technology advancement and independence, limited production runs, and the general lack of economic leverage over the industry. An exception of course, is the case where the aerospace company provides chips to other divisions in the organization in bulk quantities.

In general it appears that the creation of an in-house solid-state manufacturing facility is a questionable long-term cost-effective solution to the problem. Specifically, the economic law of supply and demand will become a dominant factor relative to the final solution. That is, if the number of firms having in-house solid-state technology R&D and manufacturing facilities increases unabatedly with time, then it follows that in turn, the aerospace firms will become contributors to the presently defined technology and manufacturing external factors over which they have little control. It is also not unrealistic to envision that with time the aerospace firms will also become suppliers of microelectronic circuitry to the marketplace and thus eventually become economic competitors with today's solid-state electronics firms. To use the cliche, the solution becomes part of the problem.

GENERALIZED ECONOMIC MODEL

Considerations, Factors, and Parameters

On many occasions the outputs obtained from computer-related economic analyses do not match the results expected. This is due not to shortcomings in the theory, but rather with the inability of the analyst to properly identify all parameters and factors which go into the analytical model itself. Additionally, it is assumed that external factors such as described earlier are considered in some fashion by management, but do not enter into the definition of the model for obvious reasons. While it is fully recognized that these external factors can have deleterious effects upon the cost of a DCCS, they are variables that do not lend themselves to analysis as part of a Generalized DCCS Economic Model. They can be perhaps more properly identified as economic factors which impact the DCCS model in terms of accuracy of the results.

The Generalized DCCS Economic Model which follows is based largely on experience to date with Avionic System Architectures which utilize a centralized data processing architecture as the main computer resource onboard the aircraft. It is the opinion of the author that this experience is directly translatable to a DCCS as pertains to what considerations, factors and parameters should be included in the model. Unfortunately, what is also being carried forth are certain inherent deficiencies in the costing of computer resources associated with today's aircraft/avionic systems. In large part, these deficiencies are software related. It should be self-evident that going to a Distributed Computer Control System will not overcome the lack of accountability and accuracy in software cost analysis.

Generalized DCCS Economic Model

The Generalized Economic Model presented herein leads to a straight-forward, additive algorithm that is best described as a sum of the elements of a three-dimensional array. Consider the following:

C_{1jk} = The material cost of a unit micromachine[1] that includes CPU, memory, and I/O costs (if all such are present).

C_{2jk} = The cost of an individual micromachine installation including the costs of labor and materials identified with that installation.

---

[1] Here micromachine refers to both microprocessor and microcomputer.

C_{3jk} = The cost of testing a micromachine including both labor and materials

C_{4jk} = The cost of the application (or function) software development for a micromachine.

C_{5jk} = The cost of software debug (test) and documentation for a unit micromachine.

C_{6jk} = The cost of the I/O controller software for a unit micromachine including control protocols and operating system if appropriate.

It should be noted that there are six (6) attributes of the basic hardware/software costs. However, any machine may be replicated for one of two reasons: namely, redundancy and system standardization. Herein, redundancy refers to the inclusion of duplicate machines within a system for the replacement of that function upon incipient failure of one of the redundant machines. System standardization refers to the requirement that certain groups of machines must be of common genre so as to minimize logistics.

This leads to an unambiguous algorithm that is readily reduced to machine computation. Keep in mind that any C_{ijk} may be zero. This is to prevent duplication of cost factors where such duplication is unwarranted (e.g., software costs for redundant machines).

$$C_a = \text{System Acquisition Cost} \quad (1)$$

$$= \sum_{k=1}^{N_t} \sum_{j=1}^{M_t} (M_{jk} + 1) \sum_{i=1}^{6} C_{ijk}$$

Where N_t = Total number of different types[2]

M_t = Total number within each type

M_{jk} = Degree of redundancy

C_{ijk} = As described above

All indicies and coefficients are finite and bounded below as indicated:

$1 \leq N_t$

$1 \leq M_t$

$0 \leq M_{jk}$

$0 \leq C_{ijk}$

---

[2]Type refers to grouping according to manufacturer's part number.

Also, there are costs other than the system acquisition cost which contribute to the total DCCS cost. These costs are essentially "life cycle" costs which more clearly reflect the total cost of owning the system. These costs are:

C_d = The system design and development costs which include "non-recurrent engineering" costs (NRE), such as reliability estimation and verifications.

C_{te} = The system test and evaluation costs which include breadboard, simulation, captive and flight test schedules.

C_o = The system operating costs which account for spares and replacements, and any updates to the initial design.

C_m = The system maintenance costs which include all labor costs involved in the test and maintenance of field items during the life of the system development such as the installation and support of software maintenance centers.

C_l = The system logistics costs which include all costs associated with distributing and warehousing systems components during the period of the system deployment.

Therefore, the resultant DCCS total cost is:

$$C_{DCCS} = C_a + C_d + C_{te} + C_o + C_m + C_l \quad (2)$$

The assumptions under which this model was generated are considered to be both realistic and direct. The accuracy of the model is bounded only by the accuracy of the estimates of the individual C_{ijk} terms.

Functional, Idealized Comparison Of An Aircraft/Avionic DCCS With An Automated Process Control Plant

Figure 6 provides a graphical representation of a functional, idealized comparison between an aircraft/avionic DCCS and an automated process control plant. As is readily self-evident, the functions performed in both instances have a one-to-one correspondence. Based on this comparison, it can be concluded that the majority of the DCCS economic considerations addressed in this paper have direct applicability to an automated process control plant.

CONCLUSIONS

A Generalized DCCS Economic Model for an Avionic System Architecture (ASA) can be postulated at this point in time. However, the validity of the results can only be taken as being theoretical as there is little

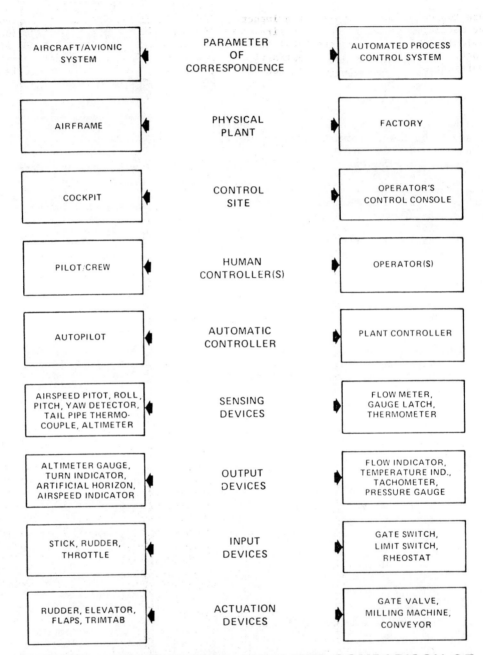

FIGURE 6: A FUNCTIONAL, IDEALIZED COMPARISON OF AIRCRAFT/AVIONIC DCCS AND AN AUTOMATED PROCESS CONTROL SYSTEM.

experience with an ASA formulated completely around a data bus-oriented distributed data processing system. On the other hand, many currently used computer resource costing techniques can be used to identify cost parameters, considerations, and factors which go into the definition of the model.

There are two major cost drivers which impact both the development and design of an aircraft/avionic DCCS as well as its total life-cycle cost. First there is the determination and specification of standardization requirements for computer-related hardware and software. Secondly, there is the impact of the technology directions and marketplace thrusts of the solid-state (microelectronics) industry over which the developer and designer have little if any control. In conclusion, the credibility, acceptability, and management worth of the Generalized DCCS Economic Model will depend on the correspondence between the parameters, considerations, and factors within the model and the aforementioned cost drivers-standardization and the solid-state industry marketplace.

AUTHOR INDEX

AlDabass, D. 143

Garcia, C. A. 21

Harrison, T. J. 75
Humphrey, J. W. 1

Johnson, T. L. 81

Kaiser, V. A. 21
Kawabata, T. 119
Kuroda, S. 53

Lalive d'Epinay, Th. 65

Miller, W. L. 83
Müller, K. D. 105

Nader, A. 157, 197

Romeu, F. J. 183

Schmidt, J. 105
Singer, R. P. 87
Steusloff, H. U. 39, 91

Tanaka, H. 119

Wilhelm, R. G. Jr. 133

Yoshii, S. 53

Zempolich, B. A. 211
Zwoll, K. 105